A Curious Field-book

A Curious Field-book
Science & Society in Canadian History

Edited by
TREVOR H. LEVERE
and
RICHARD A. JARRELL

TORONTO
Oxford University Press
1974

Cover design by
FRED HUFFMAN

ISBN-0-19-540221-9

Printed in Canada by
THE BRYANT PRESS LIMITED

Foreword

It is a pleasure to thank the many individuals who have made the preparation of this book both possible and enjoyable. Particular thanks go to the librarians at the Baldwin Room, Metropolitan Toronto Central Library and at the Rare Book Room, University of Toronto Library. The initial stimulus for the book came from Ms Tilly Crawley of the Oxford University Press. She and Mrs Martha Dzioba were most helpful in providing editorial midwifery at every stage.

Finally, acknowledgement is made to the Director General, Meteorological Office, for permission to consult and extract from The Sir Edward Sabine Collection available at the Meteorological Office Archives (Copyright Controller, Her Majesty's Stationery).

T.H.L.
R.A.J.

To Jennifer and Marti

Contents

Illustrations

All illustrations are reproduced by courtesy of the Metropolitan Toronto Central Library or the Public Archives of Canada

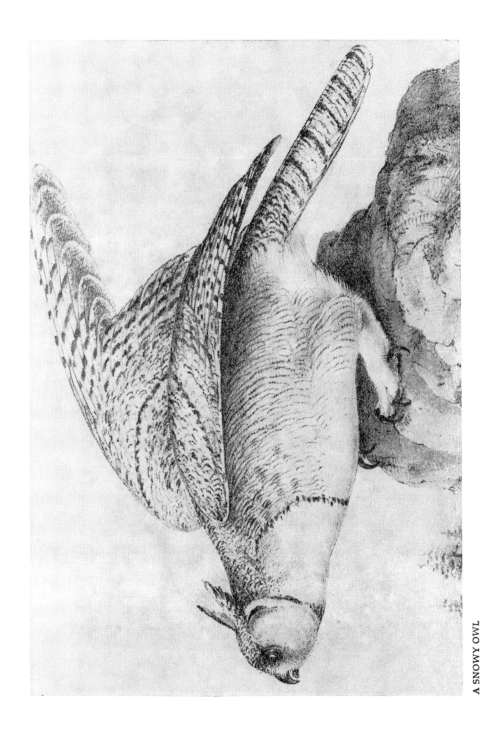

A SNOWY OWL

General Introduction

Science is tightly woven into the fabric of modern Canadian society, and plays an increasingly important role therein. The exact nature and extent of that role are now the subject of careful and extensive study and debate, yet we remain largely ignorant of the roots of our present situation. As this volume will show, these roots are complex, substantial and intriguing. A major step towards their understanding would be a systematic investigation of the history of science in Canada, a subject still in search of its author. Current research is seeking to remedy such a state of affairs. This volume, however, does not pretend to be a history of Canadian science, but rather illustrates major themes in social history, related by salient aspects of science. Historians are only beginning to probe the relations between science and the societies within which it develops, and their conclusions are therefore necessarily tentative; this study must be considered merely as a prelude, and perhaps as something of a gad-fly. Our selection of themes and extracts is naturally partial, and, for example, so important a field as the social role of medicine, worthy indeed of extensive consideration, has been almost entirely ignored in the following pages. We do, however, offer readings to illustrate some fundamental notions that are discussed in this essay, and to illuminate significant perspectives in Canadian social history. The threads that form the social history of science in Canadian history are woven into a rich if little-known tapestry that deserves and repays closer acquaintance than it has generally yet received.

Science is often described as an international discipline, unconfined by territorial boundaries or cultural distinctions. It is, when seen in this light, subsumed in the great Republic of Letters. The Republic, if it exists, is unfortunately infirm and full of dissension. Science, considered as an enterprise directed towards achieving an understanding of nature in terms of general laws, is certainly international, but when one views it as but a variable factor reciprocally interacting with many others within the social fabric, the situation is different. The nature of science and the patterns of its development are distinct from one another, and our discussion of Canadian science rests upon and implies this distinction. Modern and Romantic scholarship alike have shown that national and cultural peculiarities play a large role in the development of science in any region, creating and affecting trends in scientific education, popular skepticism, acceptance or rejection of science-based values, governmental attitudes and science policy, and even the scientist's choice of research fields.

Science in Canada is no different. Just as a uniquely Canadian ethos exists in other areas of intellectual endeavour, so is there one in science. Our initial assumption, supported by the readings that follow this essay, is that Canadian science—or, more precisely, science as it has been practised and developed in the territories and provinces that constitute contemporary Canada—has been and is unique. It was neither American nor British nor French. Although naturally owing much to those national styles of science, its terms of reference were, from the first, different from all others. Our second assumption is that the nature of Canadian science was and long remained predominantly utilitarian. This follows in part from Canada's geographical situation and the response that the environment exacted from the populace. One particular consequence is a close marriage, inimitable in style, between science and technology, inevitable in any accelerated industrial revolution in a developing country. For this reason we conceive this volume and its companion on technology in Canadian history to be particularly useful complements to one another. As we shall see, there were from the outset characteristic and persevering differences in the attitudes held towards science by English- and French-speaking Canadians. Social attitudes towards science are distinct in French and English Canada at the beginning of the colonial era, and their direct results are still with us. Although this study ends arbitrarily with the First World War, it will identify those attitudes towards science that, among French Canadians, date from the seventeenth century, and, among English Canadians, from the early nineteenth century. The roots of twentieth-century science in Canada were established by the time of the Durham report.

Plus ultra—further still!—wrote Joseph Glanvill, an original Fellow of the Royal Society of London, as the title of his popular work on the new science of the seventeenth century. That exhortation, together with the young Society's invitation to action—*Nullius in Verba* (Nothing in words) —may be taken as an epitome of the new enquiring spirit. The rise of modern science coincided and was intimately connected with the expansion of European civilization in the Renaissance and the seventeenth century. It was an age of spiritual and intellectual iconoclasm, bold uncertainty (witness, for example, Nicholas of Cusa's *Learned Ignorance* of 1440), and expanding horizons. Aristotelian views were called in doubt, there was increasing emphasis upon observation and experiment rather than ancient authority in science, and Francis Bacon's explicit recognition of experiments of fruit as well as of light, yielding material as well as intellectual benefits, led to an increasing awareness of the power of knowledge. As Renaissance natural magic gave way before natural science, many began to see science as the symbol of European power, and the expansion of knowledge as more than a literary parallel to the expansion brought about by guns and sails. The discovery of the New World—new lands, animals, plants and peoples, new wealth and new knowledge—stimulated the appetite for natural science. Bacon, in his program for the discovery and control of nature, had advocated a modification and extension of older natural histories. Careful, detailed, and systematic inventories of natural products and curiosities should be prepared. Such activities, for which Bacon was by no means the first nor the only advocate, were indeed often an important part of the exploration and colonization of the New World, so that the frontiers of empire and knowledge advanced together, particularly under the banners of England and of Spain. The northern reaches of the American continent, the future Canada, initially present an exception to this generalization. There were, of course, keen observers among the French explorers, notably Jacques Cartier and Samuel de Champlain. France, however, at first sought commercial advantages rather than colonization along the valley of the St Lawrence. The student of Canadian history well knows how precarious was the French hold thereon during the seventeenth century, and the population did not increase to the point where one might reasonably describe New France as a colony until late in the century. Fur traders, soldiers, and sailors generally had little interest in science. There was however another group—the missionaries—to whom one looks for the first scientific observations and natural histories of Canada's minerals, flora, and fauna.

The Jesuits established themselves along the St Lawrence and in the region of the Great Lakes in the 1630s, and while their main concern was naturally the conversion of the native peoples to Christianity, their intellectual vigour also flowed into natural philosophy, and they provided the bulk of scientific intelligence concerning the new land. They constituted a formidable order of the church militant, founded in 1539 and recognizing that a major cause of the Reformation was the ignorance of so many priests. To combat this the Jesuits made themselves rigorously, broadly, and highly educated, working in the world at large, mobile, cosmopolitan, and active. They were innovators in educational systems and materials, established excellent schools, and were for centuries the best schoolmasters in Europe. In Rome Jesuits were among the foremost astronomers in Galileo's day, while from Peking to New France they were zealous exponents and practitioners of natural philosophy. They were quite simply the best educated order in the seventeenth century, and in their colleges in France they pursued and provided a curriculum comprehending mathematics and science. Many of the Jesuits who came to Canada had been educated in the Collège de la Flèche, where Descartes had studied, and they brought with them a love for knowledge and a full understanding of contemporary science. Aristotelianism was still at that time basic to most European science, and the many reports returned to France annually in the *Relations of the Jesuits* frequently exhibit Aristotelian explanations of natural phenomena observed in Canada. Jesuits were trying to describe and explain the new country as early as the foundation of the *habitation* of Port-Royal in Acadia in 1604.

The Jesuits had wide scientific interests and carefully observed a variety of new plants, animals, and geological formations. They also left excellent and often sympathetic discussions of the customs, languages, and arts of Indian peoples. Besides being missionaries the Jesuits were astronomers, geologists, natural historians, and anthropologists who, in addition to enriching the old world with knowledge of the new, used European science to impress upon the Indians the superiority of European culture, including the Christian religion.

Most Canadian science in the seventeenth century was in Jesuit hands. Yet even from the founding of New France other Frenchmen had taken a scientific interest in their newest possession. Many of Cartier's observations were incorporated into European works on botany and geography and, in the seventeenth century, works solely dedicated to Canada began to appear. The first Canadian herbal, Jacques Cornut's *Canadensium plantarum historia*, was published in Paris in 1635. Although Cornut had never been to Canada, he was able to rely upon the reports of those who had. Natural history featured significantly in the *Histoire véritable et naturelle* (1664), a work of propaganda by Pierre Boucher, the governor of Trois-Rivières. A number of other works on ethnology and natural history by Marc Lescarbot, Nicolas Denys, Gabriel Sagard, and others provided Europeans with a reasonable knowledge of New France. Official France however took little interest in the natural resources and productions of the country until mid-century when the accession of Louis XIV and the able work of ministers like Richelieu, Mazarin, and Colbert opened a new phase in French colonial expansion. The most important event for Canadian science during the French régime was the founding of the Académie Royale des Sciences in Paris. From the beginning the Academicians took an interest in New France. They were fortunate in having, in the early days of the eighteenth century, an able correspondent in Quebec, Michel Sarrazin, the royal physician. For many years Sarrazin sent a stream of information to the Academicians Tournefort, Bignon, and Réaumur.

During the early eighteenth century the French received scientific information from several sources in New France, but those providing the scientific observations were from France, not Canada. The climax in scientific work in Canada came at mid-century during the temporary governorship of Roland-Michel Barrin de La Galissonière. Himself a Free Associate of the Académie, the Marquis de La Galissonière, through his Parisian friends Bouguer, Guettard, Réaumur, and Duhamel du Monceau, maintained a flow of information to French scientific circles. He ordered military officers to provide him with new discoveries and, on his recall, the flow of information continued, principally directed to Guettard by the royal physician at Quebec, Jean-François Gaulthier. The botanist Pehr [Peter] Kalm, visiting Quebec in 1749, was astonished at the interest the French took in natural history, and was pleased to find men like La Galissonière and Gaulthier in the small colony. La Galissonière in particular brought to New France the ideals of the French Enlightenment. Educated patronage, a belief in social and intellectual progress founded upon the proper application of reason and scientific method, an avid curiosity about the natural world reinforced by the conviction that knowledge thereof could not but be ultimately useful, hopes for an intimate and fruitful union of science and technology—these were major ingredients of his vision, as they were of Diderot and the other Encyclopaedists. With such a governor New France could become Bacon's New Atlantis, and the aspirations of men of science were mightily encouraged. Upon La Galissonière's recall to France such dreams vanished as surely as Atlantis itself. The values of science however were not utterly denied.

The Jesuits originated as soldiers of the Counter-Reformation, not as children of Enlightenment, but they too had their vision and their uses for science. In spite of the colony's small population they made their first attempts at scientific education even before the mid-seventeenth century. They opened the Collège de Québec in 1635—prior to the foundation of Harvard—for the higher education of both French and native youths, and provided instruction in mathematics and science shortly thereafter. The curriculum was based on that of the Jesuit colleges in Europe and was, considering the circumstances, surprisingly complete. The most important teacher in Quebec, from the scientific point of view, was the professor of hydrography (also known as the Royal Hydrographer), whose chair had been established at the instigation of the Intendant Jean Talon for the training of pilots. A succession of fine teachers occupied the post, the most famous being Louis Jolliet, co-discoverer of the Mississippi River. Among his successors were the noted cartographer Jean-Baptiste Franquelin and the versatile Joseph de Bonnécamps. A small college in Montreal was maintained by the Jesuits for several years late in the seventeenth century, and classes in mathematics and navigation were held there by Father Claude Chauchetière. There was little else besides these bright lights. Science and mathematics were almost entirely missing from primary education in the colony, as was indeed the case for most of Europe, and the only library of scientific and mathematical works was a small collection of books in the Collège de Québec. Primary education was perhaps as good in Canada as in France, but serious limitations were apparent in higher levels of education. Father de Bonnécamps had attempted to obtain astronomical instruments for the erection of an observatory in Quebec, but to no avail. There were no laboratories, libraries, museums, academies, journals, or even a printing press in the entire expanse of Canada. This situation was to continue down to the Conquest in 1759. Then, under English occupation, it immediately worsened. With the departure of the French political and military officers, and on British orders the removal of the Jesuits, the growth of science in Canada came practically to a halt. There remained only a handful of educated people with a taste for natural history, most of them being physicians and clerics. As it turned out they were to maintain the natural history tradition of New France and to provide practically the only science in French Canada until the end of the nineteenth century.

Native-born Canadians were not much disposed to the study of science. The *habitant* and *coureur de bois*, who formed the majority of the population, daily faced a rigorous existence and had neither the education, leisure, nor inclination to cultivate things scientific. French Canada before the Conquest did not produce a significant middle class nor an upper class at all, yet these were the classes that provided France with scientists and their patrons respectively. In a country like Canada no group had enough leisure to devote itself to the study of science. Gaulthier, for example, collected specimens and reported observations only in his few spare moments away from medical duties in Quebec. In addition to the economic and military exigencies of Canadian life, another factor hindered the growth of popular support and enthusiasm for science among the French Canadians: their attitude towards contemporary French thought. Science and philosophy prospered in France during the Enlightenment and the flurry of scientific activity during La Galissonière's brief governorship reflects, albeit weakly, French achievements at that time. Enlightenment in France, spread under the aegis of the *Encyclopédie* of Diderot and d'Alembert, was often associated with deism, anti-clericalism, or both. Things were otherwise in New France and Canadian educated society as a whole was less intellectually radical than that in the mother country. One powerful reason for this was the staunchly maintained and uncompromisingly wielded authority of the Church in Canada, which was redoubtedly conservative. The Catholic Church itself clearly had no policy of hostility to science in the colony. On the contrary, missionaries, the secular clergy, and even the nuns of several orders had received some scientific and mathematical education, and the Jesuits in particular, as we have seen, were the principal exponents and practitioners of science in the land. The most impressive intellects were however French, not Canadian. La Galissonière and Dupuy were from the mother country. The Jesuits were almost exclusively French born and educated, but they gradually concentrated more and more on their missionary activities. Parish priests on the other hand were, as the eighteenth century wore on, increasingly Canadian, trained at the Seminary of Quebec. As the local representatives of the Church they were the strongest conservative influence, for the Church of New France was ultramontane, owing direct allegiance to Rome and uncontaminated by the relative libertinism of the Church of France. The priests' authority tended to the maintenance of religious orthodoxy and obedience to the state, and had not been displaced by the end of the French régime in Canada. This constituted a limiting factor, inculcating in the majority of French Canadians a conservative outlook oriented towards Church and home that could scarcely stimulate intellectual boldness or foster an attitude conducive to the expansion and development of science. Two points should be stressed. First, there is a real difference here between the colony and its parent. Secondly, cultural conservatism, little suited to the growth of science, was to be of great significance in helping to ensure French-Canadian cultural and linguistic survival in a largely British continent. This situation, involving indifference, ambivalence, and sometimes even hostility towards science, arose as early as the seventeenth century, and its effects were striking by the mid-eighteenth century, and constantly recurred thereafter.

The fall of New France paved the way for an entirely different scientific tradition in Canada, British in origin but transformed into a distinctly Canadian entity. This scientific tradition was slow in coming, for the first British immigrants to Quebec after 1759 were merchants and soldiers, little likely to practise or patronize science while still engaged in subduing and exploiting their newly won territory. Within a decade and a half, Britain was at war with thirteen of its North American colonies and Canada had become both a base of operations and a theatre of war itself, unfavourable to the advance of science. After hostilities had ceased however the colony needed science to help it expand westwards and to develop the recently constituted province of Upper Canada. To open the new western region, surveys of the land and water were necessary. The British government had appointed a Surveyor-General, Samuel Holland, whose work involved some science tangentially, chiefly consisting of astronomical observations. A complete hydrographic survey of the St Lawrence valley and the Great Lakes was clearly required, although not attempted until Bayfield's work in the 1820s.

The British had however explored other parts of Canada during the preceding century. The search for a northwest passage had stimulated a number of Englishmen to send expeditions into the northern archipelago and into Hudson's Bay. A desire to control the northern fur trade had operated as an additional incentive for such voyages. James and Ellis were among those who made wide-ranging surveys of flora, fauna, geographic formations, and sea routes in the North during the seventeenth and eighteenth centuries. Yet such voyages were not part of a concentrated effort to explore and survey the northernmost parts of the continent. The celebrated expeditions of John Franklin and John Ross in the early nineteenth century were among the first attempts to obtain systematic and comprehensive information about British North America. Such works as the *Fauna Boreali-Americana* (1829) by John Richardson, naturalist to the Franklin expedition, went far to increase British awareness of northern Canada.

Further to the south, the first explorations of Canadian topography, geology, botany, and zoology were not even undertaken by the British, but by others who were sometimes subsidized by the authorities. André Michaux carried out botanical studies in Quebec, the botany of the Maritimes was surveyed by Frederick Pursh, and a masterful topographical study of Canada was written by the French-Canadian Surveyor-General of Lower Canada, Joseph Bouchette. British science in Canada scarcely existed prior to the second decade of the nineteenth century, when the first significant stirrings occurred. Among the factors underlying this change was a new pattern in immigration.

The arrival in large numbers of United Empire Loyalists and increasingly of Scots was to have lasting effects upon the Province of Canada. Scottish immigrants were to be the most important for the practice and development of science, and they feature very prominently in any list of the chief scientific minds of nineteenth-century Canada. Reasons for this are not hard to find. Scotland, not England, was the centre of British science in the late eighteenth and early nineteenth centuries. First, and generally, Scottish notions of education were wider and more democratic than English ones. As a result, the Scottish immigrant was likely to be more highly educated than his fellow Britons. Secondly, Scottish universities had long been in touch with the best continental ideas and teaching in science and medicine, had absorbed them with vigour and enthusiasm, and incorporated them into their curricula, offering first-rate medical and scientific instruction. In such respects the Universities of Edinburgh and Glasgow far outshone their counterparts of Oxford and Cambridge, which were indeed sometimes excelled in scientific education even by dissenting academies in the midlands and north of England. In addition, Scottish professors were often keenly aware of the virtues of applying science to agriculture and industry, an important consideration for success in a developing nation. The Scots, armed with the latest ideas in science, from the latest geology to the new continental ideas in physics and chemistry, found a rich field of opportunity in Canada. When the first scientific societies were formed in Lower Canada in the 1820s, Scots figured prominently in them. The establishment of scientific education in Canadian colleges and universities during the nineteenth century relied heavily upon Scotland for its model and for its supply of faculty. Scotland furnished not only professional and semi-professional scientists and engineers, but also merchants and entrepreneurs to contribute to the scientific as well as economic development and exploitation of Canada. Scots were in the forefront of the fur trade, and later of mining, road and railroad construction, and contributed to the cultural as well as the material well-being of the Province. James McGill, for example, was a fur trader who made the first large bequest to a college in Montreal, and McGill University was granted its charter in 1821, later achieving pre-eminence in medicine, science, and engineering. William Dawson, born in Nova Scotia and educated at Edinburgh, brought the school to its pre-eminence, while its success was ensured by yet another Scottish-Canadian, Sir William Macdonald. A young country as vast and potentially as rich as Canada was fortunate in having attracted so many hard-working and ambitious Scots from its very foundation as a British colony.

To the other major group, the United Empire Loyalists, Canadian science owes a different debt. Among the first members of the stream of Loyalists entering the colony were well-educated members of the upper and middle classes. From the first they pressed for higher education in the Maritime provinces of Nova Scotia and New Brunswick. Even before mid-century, in part because of their influence, fine colleges were open in both provinces, colleges that would subsequently teach science as an integral part of the curriculum. The Loyalists were the first major group to settle in Upper Canada, where they founded a society different from that of either the Maritimes or Lower Canada, and combining elements of British, Loyalist, and Republican ideals in a way that would eventually dominate not only the Province of Canada, but the style of science therein. Time was needed to form this style, so that science did not become fully accepted as an instrument of exploitation and exploration in Upper Canada until the middle of the century.

Science in the young colony had to depend upon two groups: dedicated amateurs and the government. Private patronage, already dying in Europe, could not flourish in its old form in Canada, and the stability that facilitates the benefactions of twentieth-century patrons had not yet been achieved. Any possibility of the development of traditional patronage was destroyed in the Rebellion of 1837 when the upper classes found their position in society seriously weakened. With the growing ascendancy of the middle classes in Lower and Upper Canada came an increasing emphasis upon utilitarian pursuits and financial gain, with consequences for science that will be considered below.

British North America experienced its first major expansion in population and economic strength between 1820 and 1850, a time of British advancement and, relatively, of French-Canadian introspection. These years saw the first deliberate and widespread attempts at cultivating and applying science in Canada, and also of using the fruits of science to persuade potential immigrants that they should come to Canada and share its wealth. Abraham Gesner is representative of the type of scientific figure in early nineteenth-century Canada. He carried out careful surveys of the geology of Nova Scotia and used his findings to exert pressure on the government for the proper scientific development of the region, as well as urging capitalists to invest and immigrants to flock to participate in the great riches of the land. William Dawson, later to become Canada's scientific sage and doyen, also investigated Acadian geology, while in New Brunswick James Robb of King's College, Fredericton studied botany and geology. The government asked J.F.W. Johnston to carry out a study of the agricultural potential of New Brunswick and he did so with the personal support of the Lieutenant-Governor, Edmund Walker Head, who later became Governor-General of Canada.

Lower Canada, however, was the centre of this widespread scientific activity. A number of outstanding French Canadians, notably Bouchette, the abbés Desaulniers and Duchaîne, and Jean-Baptiste Meilleur, made useful contributions. Meilleur, for example, who was a medical doctor and educationalist, wrote the first chemistry text specially written for use in French-Canadian schools, the *Cours abrégé de leçons de Chimie* of 1833. French-Canadian journalism was lively and often adverted to matters of scientific interest and amusement. Journals like Michel Bibaud's *Bibliothèque Canadienne* and its successors contain curious articles on the unfortunate physiological consequences of drinking tea, the scientific principles of artificially ageing wine by electrification, and other more or less serious and informative items of scientific fare, juxtaposed with literary and political contributions. French literary society had long been more culturally comprehensive than English, more explicitly regarding science as but a part of general culture and mixing it freely with politics, history, and poetry, but what in Europe had been to some extent merely a distinction of form and organization seemed more essential in the Province, further divorcing British and American utilitarianism from French culture in Upper and Lower Canada.

Not surprisingly, much of the early science associated with the exploration of Canada by the British was in the hands of the military. Many early students of Canadian geology and hydrography were officers of the navy, Royal Engineers, and ordnance. When the first Canadian scientific society, the Literary and Historical Society of Quebec, was founded in 1824, a number of its members were military officers, and its early *Transactions* contain geological papers by Lieut. Baddeley, R.E., Capt. Bonnycastle, Lieut. Ingall, and Lieut. Bayfield, who also provided contributions based on his survey of the Great Lakes. Lieut. E.D. Ashe, R.N., director of the Quebec Observatory, was also a member of the Society and sent papers on a variety of subjects, especially astronomy. As one might expect, the medical fraternity in Quebec provided much of the scientific interest in the Society. Their colleagues and other educated townsmen in Montreal founded the Natural History Society of that city in 1827. Upper Canada, during the same period, was practically destitute of scientists and its first scientific society, the Canadian Institute, was not founded until 1849. The Institute was initially intended to help engineers achieve professional status in the province, but it soon grew to comprehend wider scientific interests.

The principal stimulus and support for the scientific exploration of Canada in these years came not from amateur bodies but from government, both Imperial and local. There had been an attempt to establish an observatory in Canada in the 1830s but it failed through lack of local interest. The first permanent scientific institution erected in Canada was the Toronto Magnetic Observatory, built and staffed by the Royal Engineers in 1840. The Toronto Observatory was the first Canadian outpost of the world-wide network of magnetic stations established in the 1830s and 1840s on the initiative of the British government and in line with the recommendations of a committee of the Royal Society of London. The British colonial effort was headed by Edward Sabine who directed and coordinated the activities of numerous naval and military officers, maintaining close touch with his widely dispersed workers in the field. The Toronto Observatory became the centre for Canadian geophysical observations, which gradually grew in importance until they led to the establishment of the Canadian Meteorological Service. When the Imperial government began to disengage itself from Canadian internal matters in the 1850s, the Observatory's future seemed dim. Through the intervention of the Canadian Institute, the support of the Observatory was taken over by the provincial government, which converted it into the clearing-house for the growing network of meteorological stations across the country.

The Imperial government had suggested the erection of an astronomical observatory at Quebec City to give a reliable measure and determination of time for the harbour and to act as a prime meridian for longitude measurements. Built in 1849-50 and directed by Lieut. Ashe, the observatory began a useful existence. From the first it received funds from the provincial government. Within a few years the government provided financial aid to several other observatories, including one at Kingston founded in 1855 and directed by James Williamson, a professor at Queen's University and a brother-in-law to John A. Macdonald; the McGill Observatory, directed by amateur meteorologist Charles Smallwood; and later a small observatory for the port of Saint John, N.B. Smallwood was a typical man of science of the time. An English physician, he settled on Isle Jésus north of Montreal and established a private meteorological observatory in the 1840s. He published papers on a variety of subjects in Canadian journals and in Benjamin Silliman's influential *American Journal of Science*, and for several years sent meteorological reports to Archibald Hall's *British North-American Journal*. Smallwood removed his instruments to McGill University and successfully petitioned the Province for a financial subsidy. From 1860 the government-funded observatory was an important Canadian centre of meteorological, astronomical, and soil science.

The pre-eminent governmental scientific institution in the middle years of the century was the Geological Survey. A survey of the geological and mineralogical potential of the country was of the greatest importance to those who wished to exploit Canada. Several attempts to establish such a survey had failed, but in 1842 the Geological Survey of the Province of Canada was founded with William Edmond Logan as its first director. The senior scientific organization of this country, the Survey soon established an excellent and enduring reputation for itself. Logan had only a handful of assistants, including a fellow geologist, a palaeontologist, and a chemist, Thomas Sterry Hunt, American-born and educated at Yale. With only a native companion, a canoe, and a few tools, Logan or one of his associates would undertake long and arduous journeys into the largely unsettled Canadian hinterland. Perhaps even more arduous was the annual fight with the legislature for funds. Logan waged constant battle with authorities who wanted quick results, results that would end in a profit, not pure science. Logan's position was no different from that of the legion of state geologists in the United States who had to face similar problems from tight-fisted and utilitarian-minded legislatures. Considering the vastness of the land, the paucity of staff and facilities, and the often unencouraging dealings with government, the Survey's accomplishments were remarkable. Logan's achievement did more than any other single factor to awaken widespread popular enthusiasm for science throughout the Province, and also gained Canada her first truly international prestige in literary and scientific society. Canada blossomed at universal exhibitions, waxed self-congratulatory at home, and enthused over Logan's well-merited knighthood. In 1863 Logan published his masterpiece, the *Geology of Canada*, offering detailed accounts of Laurentian and Huronian systems of rocks and presenting the first thorough knowledge of the Canadian shield, source of so much mineral wealth. On Logan's retirement in 1869 his work was carried on under a series of excellent directors including A.R.C. Selwyn and George Mercer Dawson.

Not all of Canada's geology and mineralogy were explored by the Survey however. Geology was the Canadian science *par excellence* in the nineteenth century, as it was in both the United States and Britain, where geology books outsold even novels, satisfying the thirst for useful knowledge and aiding farmers, miners, and industrialists alike. Dedicated amateurs filled in many of the gaps in the picture of Canada's geology. In the Maritimes, a young, largely self-educated John William Dawson studied the rocks of New Brunswick and Nova Scotia, publishing his *Acadian Geology* in 1855. It went through several editions during the century, increasing in size to over 700 pages. The work begun by Gesner and Robb in the Maritimes was also carried forward at Fredericton by Loring Bailey, an American-born professor, and by Saint John geologists C.F. Hartt and George Matthew. These three were to be the mainstays of the Natural History Society of New Brunswick. A provincial survey of Nova Scotia was made by Henry Youle Hind, an English scientist who later became professor at Trinity College, Toronto, and headed the Assiniboine and Saskatchewan exploration of 1857-1858. This expedition, the largest hitherto mounted by the provincial government, attempted to gather scientific information about the North-West, information that would become useful when the lands passed into Canada's hands after Confederation. The Geological Survey did not take part in the early work in the West since it was limited to the Province of Canada. After Confederation the Survey expanded its work into both the East and West, and geology was firmly established in the culture and enterprise of the new nation.

Palaeontological exploration went hand-in-hand with geological work, encountering a parsimonious reaction from practical-minded legislators who came to moderate their attitude only when they found that such apparently pure science was none the less useful in the matter of discovering coal beds. Since the latter were often essential for the proper exploitation of mineral wealth, palaeontology assumed the mantle of utility, and hence of official respectability. Dawson, a persistently prominent figure, was here again foremost in the field, publishing a number of important papers and books. His discovery of the controversial dawn animal of Canada, *Eozoön Canadense*, in the Canadian Shield provided temporarily valid ammunition in debates about the origin of life and about relations between science and the Bible. *Eozoön* appeared to Dawson to be the earliest instance of life on earth, and one whose geological context strengthened the parallel between biblical and scientific accounts of creation. Unfortunately the fossil animal turned out to be merely a curious mineral formation. Dawson established himself as one of the most distinguished opponents of Darwin's views on evolution and wrote copious tracts, of great popular appeal in Canada and Britain, on the concordance between geology, palaeontology, and revelation.

Prior to governmental advocacy of the application of science to agriculture and forestry, the life sciences in mid-century Canada were less organized and co-ordinated than geology and palaeontology. The important studies carried out in botany, zoology, and entomology were mostly amateur accomplishments, although natural history was taught at almost every college and university. French-Canadian interest in this field continued, clearly manifested in the curriculum of the Seminary of Quebec and later of Laval University. Many scientifically educated clerics collected plant and animal specimens and kept diaries of their observations, in this resembling the country clerics of eighteenth- and nineteenth-century England. Their journals seem to differ from their English counterparts principally in their greater immunity from natural theology and its degenerate descendant, physical theology, which sought to derive God's wisdom and power from the contrivance of the creation. There were many distinguished students of natural history in Lower Canada, outstanding among them being the abbé Ovide Brunet, who set out to create a botanical garden in Quebec and wrote an admirable textbook on botany, and the autodidact Léon Provancher. Provancher's *La Flore Canadienne*, the first truly major botanical work by a Canadian, appeared in 1862, while his special interest in entomology led to the publication of a series of works under the title *Petite Faune Entomologique du Canada*. His most widely influential contribution to Canadian life sciences and indeed to the cultural life of French Canada was the foundation in 1868 of his journal *Le Naturaliste Canadien*.

We have stressed the amateur status of those actively pursuing scientific studies in mid-nineteenth century Canada, for the professionals, mostly employed in government posts such as the observatories or the Geological Survey, were few in number. It should, however, also be noted that this may in part reflect the status of science at this period rather than depend upon colonial recalcitrance, for significant and widespread professionalization of science did not occur in Europe until well into the second half of the century. There were however special local considerations. In general, scientists, who lacked even this corporate name until William Whewell coined it for them in 1840, needed to become organized before they could achieve recognition as professionals, and such organization in Canada came only after Confederation and the achievement of steady intellectual growth. Until then it was unrealistic, indeed almost meaningless, to talk of a national Canadian scientific society. Local groups, for example the members of the Canadian Institute in Toronto, might consider their eventual growth into a national society, but meanwhile British North America was not a nation politically, economically, or culturally. Regionalism had not decreased since the turn of the century and, with the gradual withdrawal of Imperial support, Canadian scientists could not look to a central government for financial support. The Province of Canada, richest and most populous of the British North American Colonies, gave limited support to science for utilitarian reasons. The Maritime provinces, as yet unable to provide regular support for any scientific establishment, concentrated on scientific exploration of their potential natural resources. Such activities, although often impressive, were necessarily irregular and un-co-ordinated until the new federal government began to expand its scientific institutions.

Before Canada became a nation there was, however, one respect in which Canadian scientists were provided with a national organization, for they often felt themselves to be British scientists as well as Canadian. Many among them were members of the British Association for the Advancement of Science, and it was a great and significant point of national pride when that body subsequently held its first meeting outside Britain in Montreal in 1884. The foundation of the Royal Society of Canada in 1882 marked the true maturation of a national Canadian intellectual outlook, and its *Transactions* frequently offered a platform to advocates of Canadian science rather than British science, the science of Upper or Lower Canada, or American science.

There were, none the less, many points of similarity as well as substantial differences between the development of Canadian and American science. Cut off from Imperial scientific ties by the Revolution, American scientists practically had to start again from the beginning to forge a scientific tradition. Science in the United States before the Civil War was, like its Canadian counterpart, primarily utilitarian. The two countries exhibited close parallels in their patterns of education in science, although the outstanding exception in the U.S. was the founding of Harvard and Yale, much earlier than the development of similar institutions in Canada. The westward expansion of the United States called for exploration and exploitation, and in a country that so firmly believed in Manifest Destiny this was much more easily accomplished. State geological surveys sprang up almost as soon as states were organized, although those who manned these surveys found themselves in a position similar to that of Logan and his colleagues: they were expected to provide a quick return on the dollar. Federal government expenditures on science were almost entirely for exploratory work. Because of American constitutional difficulties the federal government only slowly entered the scientific field, and organizations like the Coast Survey and the Geological Survey were slow to develop into full-scale departments. Only after a long struggle was a national observatory created. The opening of the American West, well under way by the 1850s, had a decided impact upon American science; the opening of the Canadian West, though later in time, would have a similar beneficial effect. Two aspects of American science should be noted: it was better organized and better funded than Canadian science, for the country was more populous and richer. In the face of these factors the attitude of the Canadian scientist towards American science was ambivalent. He did not normally see himself as part of a greater North American scientific community, although he often joined the American Association for the Advancement of Science and attended its meetings in the United States and in Montreal and Toronto.

Confederation ushered in a new confident and expansive era in Canadian science. The direct effects of Confederation itself upon science were few. Government scientists could now be drawn from a wider range of scientific manpower in the new provinces. The funding of scientific institutions and the employment of scientists, already a feature in the Province of Canada, were rationalized through the various ministries of the federal government, especially the Department of the Interior. The subsidization of scientific societies and schools was the affair of the provinces, being deemed educational and therefore within the provincial sphere according to the provision of the British North America Act. It was not the political act of Confederation that spurred science but the concurrent growth of the country and the vast improvement of transportation and communication.

In the 1860s the first movements towards professional consolidation began. Professionals and amateurs were sufficiently numerous to begin thinking nationally even before political unification, and succeeding years saw the foundation of the Botanical Society of Canada (1860) and the Entomological Society of Canada (1863), both functioning as national organizations. The same period saw the rise of regional groups devoted chiefly to natural science, for this area could readily be cultivated by the amateur as well as the professional, and helped to satisfy the intense mid-Victorian thirst for knowledge. The Natural History Society of Montreal was infused with new life and published its memoirs regularly in the *Canadian Naturalist and Geologist* from the 1850s (although its *Proceedings* had been issued sporadically since 1828). The Natural History Society of New Brunswick had a chequered career from the 1860s, and within a decade of its entrance into Confederation, Manitoba possessed a Historical and Scientific Society that published transactions. In Halifax Lawson joined with a group of dedicated local men in the Nova Scotia Institute of Science. The chief organization however was the growing Canadian (later Royal Canadian) Institute in Toronto, which could draw upon both the university and the observatory for professional participants. Its *Canadian Journal of Industry, Science, and Art* published much of the best Canadian science until the advent of the publications of the Royal Society of Canada. The title of the Institute's journal reflects the ideals of that organization and indeed of many other societies, all convinced advocates of the utility of science in a growing colony.

Alongside these larger groupings were a great number of strictly local societies and institutions aimed at the cultural improvement of communities. Athenaea, literary societies, and small groups of naturalists arose across Canada. There were also widespread attempts at popular education in science. Mechanics' Institutes, created to give some scientific and technical education to the labouring classes and to inculcate in them an appreciation of the virtues of self-help, were remarkably popular in Canada, where they fared better than in Britain. From the 1840s almost until the end of the century, Mechanics' Institutes or similar organizations could be relied upon for the transmission of popular science as well as for literary and social functions. Nineteenth-century cultural and literary society was not so divorced from science as it is today on both sides of the Atlantic. Cheap editions of scientific works and biographies of heroes of science enjoyed impressive sales throughout North America. In university towns scientific societies for students and the general public provided yet another outlet for the popularization and diffusion of science, itself a branch of useful knowledge.

The entry of science into popular Canadian culture on so broad a front inevitably introduced considerable debate on the relations between science and religion, one of the staple items of late nineteenth-century controversy. The implications of geology for biblical views about the age of the earth were widely argued and more strenuously refuted by Canadian scholars than by British ones. Reactions to Darwin's *Origin of Species* and, worse still, to his *Descent of Man*, were often impassioned, and sales of critical works testify to the extent of interest in so dangerous and exciting a subject. There was no lack of comprehension of Darwin's views, but much intelligent re-interpretation of his facts, the better to accord with orthodoxy—frequently, be it noted, sectarian orthodoxy. The religious reception of geological chronology and evolutionary biology often was strikingly stubborn and skeptical in Canada, and it was also confidently so. James Bovell's *Outlines of Natural Theology*, published in the same year as the *Origin of Species*, exhibits a reasoned calm almost of the eighteenth-century in its certainty. Such views were more frequently assailed as the official Churches expressed caution, and as higher scientific education became more widespread across the nation.

The principal sources and expressions of Canadian nationalism in the nineteenth century, such as John A. Macdonald's National Policy, have been well documented. The role of science in the growth of nationalism is little known but worth exploring in the light of suggestive instances. The Geological Survey was internationally known and recognized by the scientific community but pride in it as a specifically Canadian achievement came, as we have seen, not from its impressive scientific labours but from its successful displays at the Great Exhibition of 1851 in London and at the Universal Exhibition at Paris in 1855. Logan's collection of rocks and minerals was a centre of attraction and the people back home were delighted. 'Canada against the world!' loudly proclaimed one newspaper. The visits of the British and American Associations for the Advancement of Science likewise elicited national as well as local pride when Montreal and Toronto were hosts. The degree to which the findings of science were used to paint glowing pictures in handbooks for potential immigrants is symptomatic of the same pride in the strength and virtue of Canadian science, both as intellectual achievement, as specifically Canadian achievement, and always, as a key to unlock the wealth and power of a strong, youthful country. Scientists often shared this pride—the Presidential Addresses and even some scientific papers offered to the Royal Society leave no doubt on that score—and were frequently shrewd enough to manipulate it to the advantage of their researches when negotiating with government officials.

The nation's steady growth from the 1870s to the end of the century was reflected in that of science. As Canada prospered so, symbiotically, did its science. Agriculture flourished in central Canada and much of the good arable land in Ontario, Quebec, and the Maritimes was under stable cultivation. The farmer may have had little use for geology, botany, or zoology, but the fruits of agricultural science—more properly, science applied to the art of agriculture—were of great value in his work, and earnestly considered in the agricultural press. An interest in agricultural chemistry, especially in the development and use of fertilizers, goes back to the beginning of the century in Canada. As early as 1820 an abridgement and translation into French of Humphry Davy's *Elements of Agricultural Chemistry* had been made by A.G. Douglas, and there were frequent attempts to interest practical farmers in scientific agriculture. Aubin, Gesner, Robb, and many others wrote works directed at the farmer and telling him how to benefit from science. The greatest appreciation for such efforts came in Ontario, where lively articles and correspondence on agricultural chemistry could be read in the pages of the *Canadian Agriculturist*, founded in 1849. Local councils and government fostered interest in the scientific advancement of farming by setting competitions for essays on insect pests, supporting courses of lectures on agriculture, introducing the subject into the curriculum at Toronto, and eventually by the foundation of the Ontario College of Agriculture at Guelph.

The needs of the Canadian farmer, especially as the West began to attract more immigrants, were not lost on the federal government. The Central Experimental Farm was founded in 1886 with William Saunders as director. Over the next three decades Saunders, his sons, and coworkers attempted to improve many cereal grains and to adapt them to the Canadian climate. One of their greatest successes was Marquis Wheat. When the period of the wheat boom came, Canada owed a great deal to such experimental work. The needs of economic agriculture forced the government to support other areas of botany and entomology, and in the latter part of the century the federal government provided for a National Herbarium, a Dominion Entomologist, and a Dominion Chemist, whose work was primarily concerned with agriculture. Ottawa also employed a number of excellent workers in the life sciences, including the Saunders, J. Macoun, Frank Shutt, and James Fletcher.

Fishing was another staple Canadian industry whose scientific development involved government support. In 1852 Pierre Fortin had been appointed to look into the Gulf of St Lawrence fisheries, and he initiated the systematic study of fish in Canadian waters. Some studies were carried out under the aegis of the Department of Marine and Fisheries after Confederation, but only in 1897 was a Marine Biological Station established.

Fisheries research in the early twentieth century was organized under the Biological Board, the ancestor of the Fisheries Research Board. The third great staple industry, timber, received very little scientific notice from either governments or universities until the turn of the century. This was partly owing to the logging industry's conviction that an endless supply of good wood was available, even though able scientists like Dawson had for decades warned that Canada had far from inexhaustible resources, and that development rather than exploitation of these resources was needed in order to maintain healthy national growth.

The physical sciences in Canada during the latter part of the century were of relatively little social significance. The geological sciences continued to provide the most outlets for government and university science. The Geological Survey, now in the hands of G.M. Dawson, J. Tyrrell, and others of the next generation, extended the work of the department into the Rockies and the Arctic. Surveying, astronomy, and geodesy grew apace with geology in the survey of the prairies that began in the 1870s. Astronomical observations to ascertain latitude and longitude had earlier been the function of the Quebec Observatory and various surveyors, but with the westward extension of the railroad a systematic survey had to be undertaken. An Astronomical Branch was established in the Department of the Interior in the 1880s in order to survey the Railroad Belt in British Columbia. Accurate large-scale cartography, including boundary surveys, also needed astronomical observations. Surveys of the Maine-New Brunswick border early in the century, and of the forty-eighth parallel near the mid-century, had been joint British and American projects, while Canadian scientists were involved in the Alaska Boundary Survey in the 1890s. The Astronomical Branch, led by W.F. King and his able assistant O.J. Klotz, established official time for Ottawa in the last decade of the century and began lobbying for a permanent national observatory to act as a prime meridian for a complete longitude network across Canada and eventually around the globe, a basic requirement for reliable exploration, navigation, and cartography. King was fortunate in being on good terms with both Laurier and Clifford Sifton, the Minister for the Interior, and also fortunate in asking for an observatory when Canada's economy was booming. The result was the Dominion Observatory, erected on the grounds of the Central Experimental Farm in Ottawa in 1904-5. The observatory was, from its inception, internationally recognized for its work in timekeeping, geophysics, and stellar spectroscopy.

In spite of such outstanding items as Rutherford's work in Montreal on the structure of the atom, physics and chemistry did not become well established in Canadian intellectual life or professional science until the present century, and the few chemists employed by the federal government were attached to existing departments, notably agriculture and geology, for practical work. There were no specifically chemical research laboratories in either government or industry in the nineteenth century, and this should come as no surprise. As long as capital was intensively invested in development only utilitarian aspects of science could achieve recognition, and science in the service of agriculture and geology would remain the prime claimant for support. Pure science, however valuable it may turn out for industry in the long run, is necessarily a poor relation when there are pressing practical problems to be solved in the immediate development of resources, and a thrusting frontier mentality is remote from the abstractions of mathematical physics. This however is only one side of the picture, for even in the late nineteenth century, physical scientists were becoming the brahmins of science.

Their status in Canada, to some extent reflected in the arrangement and membership of the first consciously national organization of intellectuals, the Royal Society of Canada, did not correspond to their activity. The Royal Society, founded in 1882 with the support of the Marquis of Lorne, then Governor-General, was meant to be both an honorific and functional institution. Although named after the prestigious Royal Society of London, its organization was more closely aligned with that of the Institut de France, with four sections comprehending French and English language, literature, and history, the physical sciences, and the biological sciences. Only the cream of the Canadian scientific community was elected to fellowship. From the first the *Transactions* of the society assumed the position of the leading scientific journal in the country. The published papers of section III, physical science, were usually few and short, but those of section IV, geology and biology, were of high quality. The Royal Society was fairly evenly divided between scientists and men of letters, but in its early days it was dominated by the scientists. Its presidents during the last two decades of the nineteenth and the first decade of the twentieth century included the most important scientific names in Canada, among whom were William Dawson, first president of the society, T.-E. Hamel, Sterry Hunt, Daniel Wilson, J.-C. K.-Laflamme, and others.

Canada had not had any real scientific lobby until the advent of the Royal Society. By the 1890s the fellows began using the Society's annual meeting as a lever for funds and projects. They were keenly interested in the development of government fisheries research, the establishment of several new scientific departments and institutions like the Dominion Observatory, and, when the war broke out in 1914, with government science policy. In their appeals to government and the public, scientists had constituted a clearly nationalistic group since at least the mid-nineteenth century, and as Canada and its scientific community grew, they took the greatest pride in Canadian achievement. Yet they held firmly to their ties with Imperial science and were zealous in fostering Imperial scientific co-operation. Indeed, Canadian scientists had pressed for and participated in international co-operation for some time. Sir Sandford Fleming, with the full support of the Canadian Institute, had been a principal in the establishment of International time zones, the Royal Society had lobbied for an Imperial Science Union, although to little avail, and, after the turn of the century, the Dominion Observatory helped to close the international longitude network by observations across Canada and the Pacific. This was all an aspect of a larger movement towards Imperial consolidation that, already evident in the late nineteenth century, rendered Canada fervently British with the advent of the First World War. The scientific community fully participated in this movement.

The Canadian government's efforts in scientific and industrial research during the war were slight indeed for a nation moving towards an advanced stage of industrialization. Not until 1916 was a move made to organize wartime scientific effort through the naming of the Honorary Advisory Council for Scientific and Industrial Research, parent of the National Research Council. The Royal Society was keenly interested and wholeheartedly supported the work of the Council. Although the Council was still without any real secretariat at the end of the war, possessed little money, and had no research facilities of its own, the impetus towards planned and organized research that had been generated during the war did not atrophy. In the immediate post-war era the National Research Council supported university scientific research and began producing its own research in the 1930s.

The nature and social role of Canadian scientific education exerted a strong formative influence on national development, reciprocally interacting with each of the various themes sketched above. The Jesuits, in their colleges at Quebec and Montreal, had taught some science and mathematics almost from the establishment of New France. Theoretical Aristotelian science formed a part of a philosophical education and a prelude to theology, while the practical course in hydrography was intended for pilots. Father Chauchetière's classes in Montreal had the same goal. This tradition was temporarily halted by the Conquest, followed by the closing of the College and the expulsion of the Jesuits. But the library remained and so too did the educated clergy in the Seminary of Quebec. With the approval and support of Governor Murray, the Seminary continued giving the classical program. Curricula that have come down to us from as early as the 1790s show that some physics, chemistry, mathematics, and astronomy were taught in the final two years of a student's program. The Seminary was fortunate in enjoying the services of a long series of excellent teachers of science, including abbés Edmund Burke, Jérôme Demers, John Holmes, and others. Scientific instruments and museum specimens were gathered at great cost for the Seminary and its sister institutions at Ste Anne de la Pocatière, Nicolet, and St Hyacinthe. Thanks to Demers' intelligent leadership, the Seminary of Quebec provided separate chairs in several sciences by the 1830s and 1840s, and

34456

students, both French and English, came from across Canada and the United States. When Laval University was finally opened in the 1850s its first rector, Louis-Jacques Casault, a former science teacher in the Seminary, ensured the place of science in the curriculum. Many of the most distinguished French-Canadian scientists were associated with the classical colleges, and while most of them were trained at the seminaries some, like T.-E. Hamel and E. Pagé, had received a scientific education in Europe or the United States. However broad the scientific curriculum in French-Canadian schools, its aim was not to produce scientists but rather to provide a liberal education, especially for those bound for the clergy, or for professions such as that of law. This essentially amateur but broad cultural assimilation of science remained typical in Quebec until the present century, and it was not until the 1920s that the essentially German-born idea of using universities to produce research workers emerged at Laval and at the University of Montreal.

Several French Canadians, notably Charles Mondelet, had argued for the inclusion of practical science in the school curriculum early in the century, but at first little was done along these lines at primary and secondary levels of education. Then, in the 1840s and 1850s, under the first Superintendant of Public Instruction for Lower Canada, Jean-Baptiste Meilleur, himself an ardent amateur of science, teachers in both the lower schools and the classical colleges became conversant with basic scientific principles. After Confederation several successive Quebec governments tried to increase the role of scientific and technical subjects in the schools. The educationalist Chauveau tried, while Premier, to interest Laval University in developing a school of science and technology, only to meet with refusal for essentially political reasons. The university did however institute extension courses, of which the first was a series of popular lectures on practical chemistry by Dr Hubert Larue. By the late 1870s the government of Quebec had started to establish schools of 'science applied to the arts'; the best of these was the Ecole Polytechnique under the able direction of U.-E. Archambault. The Ecole, which had a fine staff and a comprehensive curriculum, later became part of Laval and eventually transferred to the University of Montreal.

English Canada in the early nineteenth century had to develop its own tradition of education, implementing a variety of systems and a corresponding variety of science curricula. The first institutions of higher education were founded by the Loyalists in the Maritimes, taking their cues from the British-American tradition. King's College, Windsor, N.S., and King's College, Fredericton, for example, practically ignored science, for they looked to classical tradition for their model. The establishment of collegiate education in English Canada was a large task and a stable and clear picture did not emerge prior to the middle of the century. Nearly all colleges and universities in Canada offered some science by 1860, but in each case it was included as part of a liberal education. No school thought seriously of training men for scientific careers. Should anyone become interested in science he would remove to Harvard, Yale, Edinburgh, the University of London, or German universities.

The arts student in the typical Canadian college at mid-century would study astronomy, mathematics, physics in the guise of natural philosophy, chemistry, botany, zoology, palaeontology, geology, physical geography, and, perhaps, some anatomy and physiology. In some schools nearly half of a student's program involved science: this was especially true in schools with Scottish-educated professors, for they had had a similar education. Museums and demonstration equipment could be found in nearly all schools, but there were no permanent teaching or research laboratories, few practical courses, and no graduate education in science. Only late in the century was there any movement towards establishing the research degrees of ph.d. and d.sc., a movement that began in Germany, spread to Britain and the United States, and came thence to Canada. By the 1890s the universities of McGill, Toronto, Dalhousie, Queen's, and New Brunswick were all engaged in developing their potential for research; all, incidentally, had engineering or medical faculties that required more than a modicum of science. The social utility of research however remained to be demonstrated, and there were accordingly difficulties in obtaining financial support for programs of research. Private philanthropy had ably aided American higher education in its drive to provide a modern scientific system, but in Canada the colleges and universities rarely received private largesse. Two notable exceptions were McGill, where the science and engineering complex was provided in the 1890s by McDonald and Redpath, and Dalhousie, which received endowments from George Munro. In the 1890s the University of Toronto also began to persuade the province of Ontario to provide funds for scientific equipment and buildings. In addition to the development of science within the universities, Ontario's educational system, under the enlightened leadership of Egerton Ryerson, established a public school network that offered a small amount of science in its curriculum. The normal schools, like those in Quebec, made certain that its teachers knew some science, particularly in its practical aspects. Such advances came more slowly outside the old Province of Canada.

The First World War marked the end of an age, indelibly marking and changing the whole fabric of society, including its science. Throughout the decades that followed, education would pay increasingly greater court to science, the federal and sometimes provincial governments would increase the size of their scientific departments, the National Research Council would grow until it became the largest public centre of scientific funding and research. The scientific exploration of Canada would also continue to expand. The effects would be dramatic, for science in Canada in 1918 was still in its infancy and the changes were to be substantial and ever more widely encroaching upon the social structure. Meanwhile in 1918 research developments were relatively minor, for Canadian science was still essentially practical, as befitted a developing nation. Societies were still largely amateur and the funding of scientific enterprise was insignificant. If English Canada had yet to awaken to the full implications of science for post-war society, as Britain and the United States had already done, French Canada adopted a different stance. In a caustic editorial entitled 'French-Canadians need not apply', Huard, editor of the *Naturaliste Canadien*, would complain that the Royal Society elected only a few token French-Canadian fellows to its scientific sections. His information about the Royal Society was accurate but ignored the lack of outstanding French-Canadian scientists. Quebec was simply not a colonial society in 1918, but, as it had long been, an entity with its own maturing culture, which happily produced many fine lawyers, physicians, writers, and politicians, but comparatively few scientists. English Canada, more closely allied with the mother country, still predominantly utilitarian and entrepreneurial, gave greater prominence to the scientific enterprise. The amalgam of the two distinct cultures, variously enriched, produced Canadian science, forged from circumstance and necessity, borrowing where it could, inventing where it must, and unique. Its history remains an important element of contemporary Canadian society and culture, worthy of fuller acquaintance.

The Science of Colonial Canada | 1

Scientific investigations in French North America were rarely the results of careful planning and co-ordination, and the majority of the earliest reports on the land and people of what was to become Canada were simply by-products of missionary activities. The earliest major missionary establishment in New France was that of the Jesuits, the best-educated religionists of France. Their *Relations* provide us with a rich and varied fare of Canadian geography, natural history, and ethnology. Yet the *Relations* were not consciously meant to be scientific documents at all, but rather reports on the advancement of Christianity in the New World. The Jesuit, schooled in science and mathematics as well as in theology and philosophy, was well suited to be an observer of the new French realm; always inquisitive, forever roaming more deeply into the Canadian hinterland, he showed more understanding of the native peoples than the traders and government officials who followed him.

The Jesuits' scientific reports display a fascinating mixture of surprise and interest in new phenomena and the adaptation of current scientific theories. One of the earliest discussions was Biard's critical examination and rejection of Champlain's theory of scurvy, the *grosse maladie* of Jacques Cartier nearly a century earlier. Biard's discussion was theoretical, but European science could often be turned to the advantage of the practising evangelist. The native peoples of Canada were impressed by the Jesuits' knowledge of natural phenomena; Father Paul Le Jeune, for example, could employ his superior knowledge of astronomy to win over a quick-witted native. Such manipulation of the Indians through science can also be seen in the *Relation* of 1674.

The French were astonished at the extremes of Canadian weather, and the better educated tried to explain it in the light of their meteorological theories inherited from Aristotle. Bressani's account of the Canadian winter is an archetypal discussion; awed by the climate's severity and pleased by the salubrious effects the winter had upon Europeans, he tried to force the facts into the traditional scientific framework. The same point of view can be seen in other Jesuit descriptions of the geography, flora, and fauna of Canada. In the last years of the Jesuit mission to Canada, one of their most accomplished scientific brethren, Father Joseph de Bonnécamps, professor of hydrography at Quebec, accompanied a military expedition to the Ohio River valley and made a series of delightful comments on the natural environment of the American midwest. Bonnécamps's *Relation* is one of the best: his critical acumen is accompanied by lively interest in his surroundings.

The last years of the French régime saw the first official attempt to develop science, initiated during the brief governorship of the Marquis de La Galissonière, a learned man in touch with some of the best-known French scientists of the day. La Galissonière saw to it that a steady stream of scientific intelligence reached his friends in Paris; he acquired some of his information personally, some by way of his friend Dr Gaulthier in Quebec, and some through his officers Chaussegros de Léry and Lotbinière. Our best source for the practice of science at Quebec in the mid-eighteenth century comes from the keen scientific observer and traveller Pehr [Peter] Kalm. Kalm, a Finnish botanist, student and disciple of Linnaeus, was dispatched to North America by the Swedish Academy in hopes of discovering plants adaptable to the Swedish climate. He was pleasantly surprised to find La Galissonière and his small group in Quebec. Jean-François Gaulthier, the *médecin du roi* and successor to Sarrazin, was one of this group, and, like his Governor, was in touch with the leading lights of the French scientific circle. Gaulthier's letters attest to the admirable but unsystematic field work he produced. He was one of the few scientifically informed laymen who helped to construct a more complete picture of the natural history of New France. The period around 1750 was a high point of French science in Canada, for the struggle with Britain for the vast American empire of France was already beginning. After La Galissonière's recall, science was replaced by preparation for war.

Scientific education under the French régime was more enduring. Mathematics and science were introduced early into Quebec by the Jesuits, for the Collège de Québec was intended from the first to be an institution like all Jesuit schools. Though small, with little money and few teachers, the Collège provided a European education to sons of Frenchmen and natives alike. Under a succession of well-trained teachers, navigation, mathematics, and astronomy were taught from the late seventeenth century until the Conquest. This practical science was dispensed alongside the academic and abstract Aristotelian science and philosophy of the times, for the Jesuits, more than any other religious order, believed in equipping their members with both theoretical and practical tools. The life of professor and missionary, at home in the towns and at home in the vast Canadian hinterland, was the lot of the Jesuit in New France. There was no incompatibility in these diverse roles whose essential unity is illustrated by the testimony of Father Claude Chauchetière, sometime teacher of mathematics in the small college in Montreal, sometime missionary in the North. It was this dedication, and the dedication of the other religious orders that came to Canada, that gave the principal impetus to the evolution of a group of schools in New France that is truly impressive in view of the extremely small population, the poverty of the colony, and the rigours of Canadian life. By the eighteenth century advanced studies could be pursued in Quebec and Montreal, technical training was available at St Joachim, and priestly education, which included mathematics and natural philosophy, was provided in the Seminary of Quebec. Much of the success of these schools can be attributed to Canada's first resident bishop, François Montmorency-Laval. It was his foundation, the Seminary of Quebec, that inherited the position of senior institution of higher learning when the Collège was closed by the English. The form of education instituted by the Seminary became the prototype for the classical colleges of the nineteenth century. Education meanwhile remained available only to the few, and there was no general system of schooling, as is evident from the pessimistic report of the Intendant Hocquart. French Canada, like the mother country, would have to wait for over a century for improvement on that front.

A. SCIENTIFIC THEORIES IN A NEW LAND

Biard's Relation (1616), *JRAD,** vol. III, pp. 51-3.

Let us recall how Jacques Quartier lost almost all his people, the first winter he passed in this country; and also how sieur de Monts lost about half of his the first winter at Ste. Croix, and the following one, which was the first at port Royal, he also experienced great loss, but not so much, and the third year still less. Likewise at Kebec, afterwards, several died the first year, and not so many the second. This collection of incidents will serve to show us the causes of sickness and of health, which we have experienced so differently. The common disease was Scurvy, which is called land disease. The limbs, thighs, and face swell; the lips decay, and great sores come out upon them; the breath is short, and is burdened with an irritating cough; the arms are discolored, and the skin covered with blotches; the whole body sinks under exhaustion and languor, and nothing can be swallowed except a little liquid. Sieur de Champlain, philosophizing upon this, attributes the cause of these diseases to the dampness inhaled by

*Jesuit Relations and Allied Documents (73 vols, Cleveland, 1896-1901).

those who plow, spade, and first live upon this ground, which has never been exposed to the sun. His statements are plausible and not without examples; but they may be confronted with the fact that sailors, who only go to the coast to fish, and not clear the land at all, nor live upon it, often fall ill of this malady, and especially the Bretons, for it seems to pick them out from all the others. Also, that we, who were well as I have said, worked a great deal in the soil and out in the open air, yet we scarcely knew what this evil was, except I myself, to a slight degree, during the second winter when I became very much bloated from fever and extreme weakness; but my gums and lips were not affected, and my illness passed off in ten or twelve days. I believe it was a great benefit to us that our dwelling was not new, and that, the space around the settlement having been cleared for a long time, we had a free and pure circulation of air. And I believe that this is principally what Champlain meant.

I have heard of others, who argued differently, and not without Logic. They believed that living inactive during a long and gloomy winter, like that of Canada, had been the cause of this disease among the new inhabitants. Of all sieur de Monts's people who wintered first at Sainte Croix, only eleven remained well. These were a jolly company of hunters, who preferred rabbit hunting, to the air of the fireside; skating on the ponds, to turning over lazily in bed; making snowballs to bring down the game, to sitting around the fire talking about Paris and its good cooks. Also, as to us who were always well at Port Royal, our poverty certainly relieved us of two great evils, that of excessive eating and drinking, and of laziness. For we always had good exercise of some kind, and on the other hand our stomachs were not overloaded. I certainly believe that this medicine was of great benefit to us.

Le Jeune's Relation (1637), *JRAD*, vol. xii, pp. 141-3.

On the 10th of January *Makheabichtichiou* asked me many questions about the phenomena of nature, such as, "whence arose the Eclipse of the moon?" When I told him that it was caused by the interposition of the earth between it and the Sun, he replied that he could hardly believe that, "Because," said he, "if this darkening of the moon were caused by the passage of the earth between it and the Sun, since this passage often occurs, one would see the moon [often] Eclipsed, which does not happen." I represented to him that, the Sky being so large as it is, and the earth being so small, this interposition did not happen as frequently as he imagined; upon seeing it represented by moving a candle around a ball, he was very well satisfied. He asked me how it was that the Sky appeared to be sometimes red, sometimes another color. I replied that the light, passing into the vapors or clouds, caused this diversity of color according to the different qualities of the clouds in which it happened to be, and thereupon I showed him a prism. "Thou dost not see," I said to him, "any color in this glass; place it before thine eyes, and thou wilt see it full of beautiful colors which will come from the light." Having held it up to his eyes and seeing a great variety of colors, he exclaimed, "You are Manitous, you Frenchmen; you know the Sky and the earth."

Relation of 1673-74, *JRAD*, vol. LVIII, pp. 181-5.

OF THE ECLIPSE OF THE MOON ON THE 21ST OF
JANUARY, 1674

For a long time previously, I had talked to our Onneiouts about this eclipse; and, at the very beginning of the new moon, I had challenged the elders and, in particular, some jugglers who claim to foretell events, to say in how many days it would occur. They all hung their heads, and were compelled to confess their ignorance. "But," I said, "are these persons, who say that they come from the sky, ignorant of what happens up there? Cannot these professional diviners even predict a thing that is revealed in nature? Are these men—who know fabulous stories so well, who relate such extraordinary things about the sun and the moon, who take these objects for divinities, and offer them tobacco to obtain success in war and in hunting—not aware when one or the other is to be eclipsed?" The more I pressed them, the more they were abashed. "Is it the moon that is just beginning that is to be eclipsed?" they asked me. "Yes," I replied, "it is this moon; the only question is to know when the eclipse will occur. Take courage, consult among yourselves, and let us see for a moment how accurate your art is in predicting future events." The poor people admitted to me that it was beyond their knowledge, and begged me to go and notify them at the time of the eclipse. After this avowal of their ignorance had been several times reiterated, I publicly announced on Sunday, after mass, that the eclipse would take place on the following night; and that, if they awoke, they must remember to look. Fortunately, the sky was very clear; and, as soon as I noticed that the eclipse was beginning, I went to the orator of the country, and to some others among the most notable men; they arose and, coming quickly out of their cabins, saw that the eclipse was already very perceptible. Immediately, they announced the event, within and without the fort. I warned them that it would not remain as it was; that it would increase a great deal more, and that barely one-twelfth of the moon would remain visible. They asked me whether it would not reappear again, for these simple people thought that it was almost lost. "It will reappear entirely," I said, "and then it will be at such a spot in the sky for it continues to advance; and, just as you now see it gradually growing smaller, so will you see it *grow larger* in the same proportion."

Everything happened as I had announced, and they were compelled to admit that we knew things better than they. For my part, I derived great benefit from this, in instructing them and undeceiving them about their myths and superstitions. Such perceptible things have a much greater effect on their rude minds than would all the reasoning that could be brought to bear upon them.

Bressani's Relation (1653), *JRAD*, vol. xxxviii, pp. 221-7.

SITUATION AND DISCOVERY OF NEW FRANCE

By new France is commonly understood the space of land and water which extends from 36 degrees of latitude, which is that of Virginia, to 52, where, nearly, begins the great River of Saint Lawrence; others locate it from 32 to 54. It extends in longitude from 325 degrees to 295, as known to us,—or, to speak more properly, without any limit toward the West. It is a part of the Mainland of North America, distant from Europe, in a direct course, about three thousand miles, as we have observed in various Eclipses; situated, as is seen, in one of the temperate Zones, but partaking of the quality of the two extremes,—having severe cold in Winter, very deep snows, and very hard ice; and in Summer, no less heat than that of Italy.

The first French who lived there believed that the cause of such excessive cold (which, among other things, for nearly four months renders it impossible to write, unless one ply his pen very close to the fire, to such a degree does every liquid freeze) was the endlessly vast woods which cover the whole country. But I myself believe that if the woods, dry and leafless as they are in Winter, could hinder the Sun from warming the earth and moderating the excessive cold, they would avail still more in keeping off the heat in Summer, when they are very dense; and yet they do not,—the heat in the woods themselves being then very intense, although some nights it freezes as in Winter. I think, therefore, that the true reason is the dryness, called by Aristotle the *cos caloris et frigoris* ['stimulus' of heat and cold]. I do not dispute whether the cold of new France is more intense than that of Countries which are under the same latitude; certain it is, that it is much more acute, and accompanied with much snow and ice, which keep the rivers frozen five and six entire months. But all this may be an effect of the dryness, which is necessary for the snows and ice,—it being a very well-founded opinion that even very intense cold is not sufficient to make ice; otherwise, water—which naturally never freezes except under the greatest cold, as many will have it, or at least under a highly intense cold, as no one denies—would in its natural state be frozen, contrary to its destined use, which is to serve for washing, and as a drink for men and animals. But, because cold alone, although intense, is not sufficient without either some little body, or exhalation, or dry quality, therefore water, even in its natural state, would be fluid; and where dryness prevails, although the cold is not greater than elsewhere, it contracts or expands itself into snow and into ice. Besides, the dryness of these countries is evident,—first, because most of the lands are either stony or sandy (but not, on that account, sterile), whence the Sun cannot derive other than very dry exhalations; and the maritime countries, as being more moist, have less snow, and it melts more quickly. Secondly, from experience, through the scarcity of rains, and by the salubrity of the air, so great that, in sixteen and more years during which the Huron Mission has lasted,—where, during the same time, we have been as many as sixty Europeans, among whom were many of very feeble constitution,—no one has died a natural death here, notwithstanding the great inconveniences and sufferings, as we shall see; while in Europe those years are few indeed when some one does not die in our Colleges, if their inmates are at all numerous. Now, *omnis corruptio ab humido* [all corruption comes from what is humid],— therefore, *à contrario, sanitas à sicco* [contrariwise, health comes from what is dry]; and on this account, perhaps,—besides the change of diet,— the Barbarians find it difficult to accustom themselves to the air of Europe. Thus there is a common cause for both heat and cold, namely, *quia siccitas est cos caloris et frigoris* [because dryness is the 'stimulus' of heat and cold]. But for the cold, in particular, we might add: First, that the land lies higher than ours, and consequently nearer to the second region of the air, of whose cold it partakes in a greater degree. And this is proved by the greater depth of the Sea, which is consequently more dangerous to the ships that are obliged to land. Secondly, by the many river-cascades, which if placed together would form a fairly high mountain; which, however, forming itself, as it were, by gradations, is not so perceptible. Thirdly, by the very cold winds blowing from the neighboring mountains, which traverse the whole country as the Apennines traverse Italy; these winds more frequently blow from cold and dry countries, corresponding to our northwest winds, and to the Southwest wind which in those countries is cold, clear, and healthful,—the rains proceeding from the Northeast wind, which comes from the sea. The country, it is true, is full of great rivers and immense lakes; but this does not detract from its dryness,—these rivers and lakes being of very pure and very wholesome water; secondly, the bottom is of rock or sand; thirdly, they are in continual motion through the flow and ebb of the tide, whose action extends five hundred miles inland, and, finally, through the winds, which agitate them like the Sea, and thereby restrain the action of the Sun which otherwise would draw from them a greater abundance of vapors. This last is the very reason why it does not continually rain on the sea,—whose water, on the other hand, is much warmer, of greater volume, and more open to receive the influence of the Sun. Some one might add to this the nearness or contiguity of the

Seas of Canada to the icy sea,—from which, or at least from whose shores, are detached whole mountains of ice, which, in the months of June and July, are encountered even in the gulf of Saint Lawrence. I have repeatedly seen them as great as entire Cities; and Pilots worthy of credence say that they have seen some, along which they have coasted for 200 miles and over.

Père de Bonnécamps's Journal (1750), *JRAD*, vol. LXIX, pp. 157-9, 161, 165-7, 171-3, 189.

We encamped at the little rapid at the entrance of lake Erie. The channel which furnishes communication between the two lakes is about 9 leagues in length. Two leagues above the fort, the portage begins. There are three hills to climb, almost in succession. The 3rd is extraordinarily high and steep; it is, at its summit, at least 300 feet above the level of the water. If I had had my graphometer, I could have ascertained its exact height; but I had left that instrument at the fort, for fear that some accident might happen to it during the rest of the voyage. When the top of this last hill is reached, there is a level road to the other end of the portage; the road is broad, fine, and smooth. The famous waterfall of Niagara is very nearly equidistant from the two lakes. It is formed by a rock cleft vertically, and is 133 feet, according to my measurement, which I believe to be exact. Its figure is a half-ellipse, divided near the middle by a little island. The width of the fall is perhaps three-eighths of a league. The water falls in foam over the length of the rock, and is received in a large basin, over which hangs a continual mist. . . .

. . . Lake Erie is not deep; Its waters have neither the transparency nor the coolness of those of lake Ontario. It is at this lake that I saw for the first time the wild turkeys; they differ in no way from our domestic turkeys.

The 17th. We began the portage, and made a good league that day. I observed the latitude at the 2nd station,—that is, half a league from the lake,—and I found it 42° 33'. The 18th. Our people being fatigued, we shortened the intervals between the stations, and we hardly made more than half a league. The 19th. Bad weather did not allow us to advance far; nevertheless we gained ground every day, and, the 22nd, the portage was entirely accomplished.

In my judgment, it is three and a half leagues. The road is passably good. The wood through which it is cut resembles our forests in France. The beech, the ash, the elm, the red and white oak—these trees compose the greater part of it. A species of trees is found there, which has no other name than that of "the unknown tree." Its trunk is high, erect, and almost without branches to the top. It has a light, soft wood, which is used for making pirogues, and is good for that alone. Eyes more trained than ours, would, perhaps, have made discoveries which would have pleased the taste of arborists. . . .

Monsieur Chabert on that day caught seven rattlesnakes, which were the first that I had seen. This snake differs in no way from others, except that its tail is terminated by seven or eight little scales, fitting one into another, which make a sort of clicking sound when the creature moves or shakes itself. Some have yellowish spots scattered over a brown background, and others are entirely brown, or almost black.

There are, I am told, very large ones. None of those which I have seen exceed[s] 4 feet. The bite is fatal. It is said that washing the wound which has been received, with saliva mixed with a little sea-salt, is a sovereign remedy. We have not had, thank God, any occasion to put this antidote to the test. I have been told a thousand marvelous things about this reptile; among others, that the Squirrel, upon perceiving a rattlesnake, immediately becomes greatly agitated; and, at the end of a certain period of time,—drawn, as it were, by an invincible attraction,—approaches it, even throwing itself into the jaws of the serpent. I have read a statement similar to this report in philosophic transactions; but I do not give it credence, for all that. . . .

One of our Officers showed me a bean-tree. This is a tree of medium size whose trunk and branches are armed with thorns three or four inches long, and two or three lines thick at the base. The interior of these thorns is filled with pulp. The fruit is a sort of little bean, enclosed in a pod about a foot long, an inch wide, and of a reddish color somewhat mingled with green. There are five or six beans in each pod. The same day, we dined in a hollow cotton-tree, in which 29 men could be ranged side by side. This tree is not rare in those regions; it grows on the river-banks and in marshy places. It attains a great height and has many branches. Its bark is seamed and rough like shagreen. The wood is hard, brittle, and apt to decay; I do not believe that I have seen two of these trees that were not hollow. Its leaves are large and thickly set; its fruit is of the size of a hazelnut, enveloped in down; the whole resembling an apple, exactly spherical, and about an inch in diameter.

Now that I am on the subject of trees, I will tell you something of the assimine-tree, and of that which is called the lentil-tree. The 1st is a shrub, the fruit of which is oval in shape, and a little larger than a bustard's egg; its substance is white and spongy, and becomes yellow when the fruit is ripe. It contains two or three kernels, large and flat like the garden bean. They have each their special cell. The fruits grow ordinarily in pairs, and are suspended on the same stalk. The French have given it a name which is not very refined, *Testiculi asini*. This is a delicate morsel for the savages and the Canadians; as for me, I have found it of an unendurable insipidity. The one which I call the lentil-tree is a tree of ordinary size; the leaf is short, oblong, and serrated all around. Its fruit much resembles our lentils. It is enclosed in pods, which grow in large, thick tufts at the extremities of the branches. But it is time to resume our course. . . .

It was while with the Miamis that I learned that we had, a little before entering rivière à la Roche, passed within two or three leagues of the famous salt-springs where are the skeletons of immense animals. This news greatly chagrined me; and I could hardly forgive myself for having missed this discovery. It was the more curious that I should have done this on my journey, and I would have been proud if I could have given you the details of it.

B. A BURGEONING SCIENTIFIC OUTLOOK

Pehr Kalm's Journal, reprinted from *The America of 1750, Peter Kalm's Travels in North America* (New York, Dover Publications, Inc., 1966), vol. II, pp. 504-6.

SEPTEMBER THE 11TH

The Marquis de la Galissonnière is one of the three noblemen, who, above all others, have gained high esteem with the French admiralty in the last war. The three are the Marquis de la Galissonnière, de la Jonquière, and de l'Etenduere. The first of these was now above fifty years of age, of a low stature, and somewhat hump-backed, but of a very agreeable appearance. He had been here for some time as governor-general, and was soon going back to France. I have already mentioned something concerning this nobleman; but when I think of his many great qualities, I can never give him a sufficient encomium. He has a surprising knowledge in all branches of science, and especially in natural history, in which he is so well versed that when he began to speak with me about it I imagined I saw our great Linné under a new form. When he spoke of the use of natural history, of the method of learning, and employing it to raise the state of a country, I was astonished to see him take his reasons from politics, as well as natural philosophy, mathematics and other sciences. I own that my conversation with this nobleman was very instructive to me; and I always drew a deal of useful knowledge from it. He told me several ways of employing natural history to the purposes of politics, the science of govern-

M. DE LA GALISSONNIÈRE

ment, and to make a country powerful in order to weaken its envious neighbors. Never has natural history had a greater promoter in this country; and it is very doubtful whether it will ever have his equal here. As soon as he got the place of governor-general, he began to take those measures for getting information in natural history which I have mentioned before. When he saw people who had for some time been in a settled place of the country, especially in the more remote parts, or had travelled in those parts, he always questioned them about the trees, plants, earths, stones, ores, animals, etc. of the place. He likewise inquired what use the inhabitants made of these things; in what state their husbandry was; what lakes, rivers, and passages there were; and a number of other particulars. Those who seemed to have clearer notions than the rest were obliged to give him circumstantial descriptions of what they had seen. He himself wrote down all the accounts he received; and by this great application, so uncommon among persons of his rank, he soon acquired a knowledge of the most distant parts of America. The priests, commandants of forts, and of several distant places were often surprised by his questions, and wondered at his knowledge, when they came to Quebec to pay their visits to him; for he often told them that near such a mountain or on such a shore, etc. where they often went hunting, there were some particular plants, trees, soils, etc. for he had gotten a knowledge of those things before. Hence it happened, that some of the inhabitants believed he had a preternatural knowledge of things, as he was able to mention all the curiosities of places, sometimes nearly two hundred Swedish miles from Quebec, though he had never been there himself, and though the others, on the other hand, had lived there for years. A person who did not know this gentleman well enough would have considered him dry and only moderately pleasant, in social relations, especially for one who had not penetrated into the sciences. But the more one became acquainted with him the better his good qualities appeared and the greater became the cause for respecting a person who was characterized by everything big. Never was there a better statesman than he; and nobody could take better measures or choose more proper means for improving a country and increasing its welfare. Canada was hardly acquainted with the treasure it possessed in the person of this nobleman, when it lost him again. The king wanted his services at home and could not leave him so far off. He was going to France with a collection of natural curiosities, and a quantity of young trees and plants, in boxes full of earth. I cannot describe all the favors he showed me. It was greater than I could have expected in my own fatherland. I do not know whether the natives or the sciences will miss him most, because he was the tenderest of fathers for both, and for the latter the biggest patron and promoter that any place has been able to show. Happy the country that has such a chief! There it is not necessary to lament about egoistic and imaginary obstacles for promoting deeds of public welfare. Such a chief gives encouragement to all things that benefit a fatherland.

Paris. Archives du Canada : Bibliothèque du Museum : Manuscrit 293. P.A.C. Copy (Translation).

J.-F. GAULTIER TO GUETTARD

21 October 1752

My dear Sir,

I am delighted that you are pleased by the few minerals that I sent you last year. The good use that you are making of them is a powerful motive for me, and I pledge myself to continue such consignments as long as I am able to obtain them from all parts of Canada. It is certain that Canada contains plenty of rarities and things of interest with regard to natural history. It is a completely new country, from which one might say that nothing has yet been taken; because we have never had any governor or intendant who wished to take an interest in these sorts of researches. M. Le Marquis de La Galissonnière is the only one who had started to put things on a good footing. Canada has suffered a very great loss, in losing him. The vast extent of his knowledge, joined to his great love of the common weal, and of everything that can be useful to the state, would have brought him to the solid establishment of a colony that is almost new-born, and nothing has been done for a hundred and fifty years. What is certain is that what has been done here is due to chance and to a certain routine rather than to the wisdom of the government. We have just lost M. de la Jonquière. Never was a man so little missed, and not without reason. He has been replaced by M. Le Marquis du Quesne, a worthy pupil of M. Le Marquis de La Galissonnière, who tries in all things to walk in the footsteps of that worthy commodore. The love of the common weal, and all that can contribute to the advantage of this colony, seem to be the only motives and the only rule of his conduct, so that he is universally loved, esteemed and honoured by everyone. In spite of all that, M. Le Marquis de La Galissonnière will still always be missed because he went into an infinity of details that pleased the inhabitants of this colony.

It gave me true satisfaction to read your paper on the minerals of Canada, as much because it is assuredly well written as because it contains really interesting things. . . . Your paper reached me too late for me to add the remarks that you wanted, and to have the specimens of various minerals that you wanted from different places. So that will be for next year, which will give me the opportunity of keeping a copy of this paper and of making a small collection that may please you. I am counting on providing you with a lot more proofs that will show that Canada is indeed a country containing mines of different kinds, and I shall accompany them with specimens that may put you in the way of producing these proofs. I am very annoyed at the delay that this is causing you, but it is not possible for me to do otherwise. You will then have a specimen of the different earths, and perhaps a description of the best known places that are at present the most interesting. As for the seeds, plants and shrubs that you wanted, you will have to address yourself to M. Le Marquis de La Galissonnière, who will esteem it a great pleasure to acquaint you with everything that I send him. He likes to be obliging, especially towards persons of merit like yourself, Sir, and persons of distinguished rank like the new Mgr. Le Duc d'Orleans; there indeed was a great prince who, knowing all your merit as he knew it, would not have failed to have helped you in making a brilliant fortune for yourself. I wish with all my heart that the new M. Le Duc d'Orleans may have the same consideration and the same affection for you, and I have no doubt at all that this is the case, since the King and all great men have a taste for all that is called natural history. M. Le Marquis de La Galissonnière will still be able to convey some seeds and shrubs to the King. I am sending a good quantity this year, and I shall have the same concern next year. . . .

Since you frequently have occasion to see Messrs. de La Galissonnière, du Hamel, and Bernard de Jussieu, I beg you to tell them that many of the plants whose seeds I am sending should be sown in moss, because that is where one constantly finds these plants. I shall be careful to indicate this on each label.

You will oblige me by informing me, as far as your business allows you, if the seeds and shrubs, that have been given to the King and to M. Le Duc d'Ayen, have succeeded, and above all if the pines and fir trees can grow again and vegetate. Tell me too whether . . . white cedars take root in France when they are planted there. How does one go about taking seeds from all the resinous trees?

We have had such a hot summer that most of the coniferous and resinous trees have produced no seeds at all.

The crop of grains has been beautiful and magnificent, in spite of the drought. The stalks have been short, but the sheaves have produced fine seed.

My post has not achieved what I would have hoped and the Eskimos have ransacked it for me again, but in spite of that there has been no loss. I am counting on continuing to have it exploited.

I have the honour to be, with much affection and sincere friendship,
          Sir
            Your most humble and obedient servant
                        Gaultier
Quebec, 21 October 1752.

Bibliothèque du Museum National d'Histoire Naturelle, MS 293, Pièce 5. P.A.C. Copy (Translation).

J.-F. GAULTIER TO J.-E. GUETTARD

Québec, 2 November 1755

This year you will receive a small collection of different minerals; it would have been larger but for a great misfortune that occurred in Québec on the 7th of June last. The hospital of this town was completely consumed by flames in less than two hours' time, and nothing was saved. In the building, I had a small cabinet in which I kept all my natural history collections. Everything was burned and lost. You have lost there, Sir, a beautiful collection of minerals, and M. de Réaumur also lost a numerous collection of fishes and birds. You will have to console one another. I am going to work to repair it as well as I can, and I hope that M. de Réaumur will have reason to be happy next year. I am writing so that you may understand my excuses, which are only too legitimate, and to prepare you for the loss.

Claude Chauchetière to his brother (1694), *JRAD*, vol. LXIV, pp. 139, 149.

We are on the 45th parallel of latitude, as is Limoges, according to the computation of Clavius,—who can be mistaken only as to minutes, because the meridian star still approaches the pole, and the sun's apogee is at present in the scorpion. I know not what will become of me. As our college of villemarie is not endowed, we are not of opinion that a teacher should be maintained there any longer. We teach, however; and I am preparing myself to continue my mathematics. I have two or three of my pupils on the ships, and one is second pilot on board a King's ship. Nevertheless, our Reverend Father Superior always tells me to hold myself ready to go to the iroquois, if peace is made; or to go to Hudson's bay. . . .

. . . I am here like a bird on a branch, ready to take flight at any moment. I was very nearly going to Hudson's Bay, where the last chaplain was killed by a wretched frenchman who was in a transport of rage. It was also intended that I should go up to Missilimakinac, to assume the direction of the Huron mission. Finally, I remained here, where we have a sort of college, which is not endowed; but I think that the Gentlemen of Villemarie will not have it long unless they endow it, because the revenues of our mission are very slight. I have pupils who are good fifth-class scholars; but I have others with beards on their chins, to whom I teach navigation, fortification, and other mathematical subjects. One of my pupils is pilot on the ship which sails to the north.

Letter by [Father Joseph] Germain (1711), *JRAD*, vol. LXVI, pp. 209-11.

As regards the Quebec college, everything exists or is Done there as in our colleges in Europe—and perhaps with greater regularity, exactness, and Fruit than in many of our colleges in France. Classes are taught here in grammar, the humanities, rhetoric, mathematics,* Philosophy, and Theology. The Pupils, although less numerous than in the large towns of Europe, nevertheless possess well-formed bodies and well-regulated minds; they are very industrious, Very docile, and capable of Making great progress in the study of letters and of virtue. I speak not of the savage children, whom our fathers educate in our missions; they likewise are not wanting in cleverness, and fail not to serve God well in their own manner of speaking and of living, according to their custom. But I refer to the French Children born in Canada, who speak the same language, who wear the same kind of clothes, and who follow the same studies as those in Paris. I say that they are very intelligent, having excellent dispositions, and are capable of succeeding well in everything that we can teach them.

Report on Canadian Education attributed to Gilles Hocquart, Intendant, c. 1737. Translation of document printed in G. Frégault and M. Trudel, *Histoire du Canada par les textes, Tome I (1534-1854)*, rev. ed. (Fides, Montreal, 1963), p. 80.

The education received by most of the children of officers and gentlemen is very limited. The children scarcely know how to read and write, and are ignorant of the first elements of geography and history; it is to be wished that they were better instructed. The Professor of Hydrography at Quebec is so involved with his post as Principal of the College, as well as his obligations as a missionary, that he can only attend to the bare essentials of his responsibilities as professor.

At Montreal, the youth are deprived of education altogether; the children go to the public schools that have been established at the Seminary

---

*In 1676 Enjalran tells us the College employed a layman "who teaches all the mathematics necessary for this country. He has instructed most of the captains who bring vessels to this country".

of St. Sulpice and at the Monastery of the Charron Brothers, where they learn only the first elements of grammar. Without further help, youths can never become useful men. I think that if His Majesty would provide well for a master in both Quebec and Montreal, to teach geometry, fortifications, and geography to the cadets among the troops, and if these cadets were assiduous in their studies, then we should see subjects capable of rendering good service.

Canadians commonly have the wits, and I believe that the proposed establishment would succeed as one might wish.

2 | Government, Science, and Exploitation: The Utility of Science

The Geological Survey was the greatest single scientific enterprise in nineteenth-century Canada, the most dramatic, the most productive, the most widely known, and the only consistently if uncertainly supported project. For a variety of reasons, some peculiarly Canadian, others more generally arising from facets of Victorian culture, the Geological Survey enjoyed a unique position. First, the early nineteenth century was the heroic age of geology, rich in investigation, debate, and controversy. The major issues between Vulcanists, who held that fire was the prime geological agent, and Neptunists, who favoured water in this role, were hammered out in the first decades of the century. Catastrophism, which accounted for discontinuities in the record of the rocks by postulating a succession of cataclysmic upheavals (often including the biblical flood), gave way to the uniformitarianism of Hutton and Lyell, seeing only the uninterrupted operation of uniform laws of nature with no vestiges of a beginning and no prospect of an end. Then too Scotland, and especially Edinburgh, was both one of the liveliest centres of such debate and the intellectual cradle of many of Canada's leading geologists.

Secondly, the early nineteenth century saw the survey perfected as a particularly valid and characteristic form of science. Surveys enabled one to determine how different kinds of animal or different geological formations were distributed, and how they varied across their geographical range. Studies of such patterns of distribution and concomitant variation in the life sciences and in geology feature prominently in the history of early nineteenth-century science. The scope of Humboldt's *Cosmos* and of his researches during his travels in South America, Darwin's observations as the naturalist on H.M.S. *Beagle*, the construction of stellar maps and international magnetic charts all reflect new patterns of observation and a new form of science. Britain, as a far-flung Imperial power, was particularly well situated to pursue and co-ordinate such work.

Thirdly, and not to be under-estimated in its effects, was the enormous thirst for factual knowledge in Victorian culture, satisfied by a large periodical literature. Geology, although bearing a complex intellectual superstructure, was constituted of a plethora of facts that was, moreover, both accessible to the populace at large and also susceptible of increase by even a minimally informed observer. Geology, in short, was popular science.

It was also presented as unquestionably useful science, and thus avidly welcomed. The rise of the Mechanics' Institutes and such publications as Dionysius Lardner's library of useful knowledge came from an appetite for knowledge and faith in its power to improve material conditions, and with them the state of every level in society. Now among Canada's principal features, in the eyes of the Imperial government, of the intending immigrant, and of the potential investor of capital, was the vast potential mineral wealth of hitherto unexplored regions. A Geological Survey was greatly to be desired from every standpoint, intellectual and material. In 1859 a leader in the *Ottawa Citizen* declaimed: 'Without a Geological Survey, no Political Economist can direct the industry of Canada, or say what should or should not be done.' The extension of national resources and the advancement of abstract science often received homage as equally honourable estimates, and were seen as interdependent. As William Logan, the Survey's first director, assured the legislature, 'economics leads to science and science to economics'.

The legislature doubtless believed that intellectual vigour was desirable but regarded the achievement of economic health as a necessary prelude. From the first systematic observations in 1842 the Geological Survey of the Province of Canada enjoyed uneven relations with the government that supported it, going through an annually recurring fight for adequate funds. Logan was however eminently persuasive, the fight was increasingly successful, and the Survey grew and flourished.

Awareness of the potential utility of geological science long preceded the official Canadian survey. An article 'On the Utility and Objects of Geology' from the *Canadian Review* of 1825 spelt out what geology had already achieved in British North America, and suggested areas for future exploration. It was inevitable that this exploration would be motivated, at the official level, by primarily utilitarian aims, and the Durham report of 1839 exemplified an exploitative approach, indicative of the degree to which English interests were involved in the development of Canadian resources. A more culturally balanced approach emerged later when geology was added to the liberal arts curriculum for intellectual as well as material enrichment.

Conflict between liberal and utilitarian viewpoints bedevilled the early years of the Geological Survey. For all Logan's persuasive advocacy of the symbiosis between science and economics, the search for mineral wealth and the desire for scientific geological and palaeontological knowledge reflected different goals, apparent in Logan's personal correspondence with colleagues, especially with his assistant Murray, and with Sir Henry de la Beche, director of the British Geological Survey. Official

MINERAL LANDS.

WANTED

A Partner with Capital

To assist in the working and development of some very
valuable

SELECTED MINERAL LANDS,

BOTH IN

UPPER AND LOWER CANADA

As well as for the purchase of others of great value, in
both Provinces. Also, for the purchase and develope-
ment of large tracts of the finest Pine and other

TIMBERED LANDS,

(INCLUDING WHITEWOOD,)

In the most favourably available localities for manufacturing and shipment in Canada.

Long professional experience through the wilds of
Canada enables the advertiser to say that he is possessed of a vast amount of information respecting its
resources, in a commercial point of view, and that he
will be willing to treat with parties seeking investments on the most liberal terms.

HENRY WHITE, P.L.S.,

Author of the "Geology, Oil Fields, and Minerals of
Canada West," "The Gold Regions of Canada,"
&c., &c., &c.

intercourse with the legislature was necessarily carried on in a manner that minimized the divergence of these goals.

One major goal that the legislature set for the Survey was to determine the distribution of economically valuable metals. Two of these in particular, coal and gold, aroused widespread public interest. The utility of coal for mining and for emergent industry is self-evident, and false rumours of rich seams of coal were constantly agitated. The Bowmanville scandal was one among many. Logan however showed that there were unlikely to be extensive deposits in the Province of Canada. He did discover coal in the Maritimes, but in such thin or inaccessible seams as to be economically unpromising. As for gold, the rushes in California and Australia at mid-century stimulated the search for gold in Canada, although the *Canadian Journal* was not alone in its fear, expressed in an editorial of 1853, that gold fever was a social menace that required careful control. Gold was nowhere in the Province of Canada a common metal, and accordingly scarcely featured in the Geological Survey's initial broad delineation of the geological structure of the Province, becoming geologically significant only with the inception of more detailed work in the next decade. Economically, Canadian gold remained unimportant until the turn of the century.

In spite of battles and setbacks, and a failure to provide an instant route to riches, the Survey quickly yielded impressive results, Logan's struggle for funds became ever more successful (although never adequately so), and in 1845 he felt so committed to the development of Canadian geology that he refused a very handsome offer to direct the new Geological Survey of India.

Logan's display of Canadian minerals at the Great Exhibition of 1851 attracted a good deal of praise and roused enthusiasm at home for the work of the Survey. The growth in public awareness was matched by increasing scrutiny on the part of the provincial legislature, which established a select committee to investigate the work of Logan and his collaborators. The Report of that committee, part of which is printed in this section, is revealing both of the utility and of the politics of science in mid-Victorian Canada.

One question raised by the Select Committee concerned the possibility of an exhibit for the forthcoming Universal Exhibition at Paris in 1855. Logan agreed and, aided by the chemist to the Survey, Thomas Sterry Hunt, produced a display that brought international recognition to the Survey and aroused real pride at home. One indirect consequence of this increased fame was the unauthorized ascription of the Survey's support to promotional mining ventures. The geological promoter, who tried to persuade the public to invest capital in the exploitation of mineral resources, became a prominent figure on the scene. One of the most prolific was Henry White, a provincial land surveyor in Ontario. The illustration on page 44 suggests other motives for his work than those stated in the preface of his *Gold: How and Where to Find It!* (Toronto, 1867). The Survey had truly come of age now that its prestige had risen to the point where it was subject to piracy for private gain.

Practical science in nineteenth-century Canada was not exclusively within the province of the Geological Survey. Three other areas should be briefly mentioned: the magnetic survey, meteorology, and astronomy. The Magnetic Survey was the senior scientific institution in Canada, and in 1840 the Toronto Magnetic Observatory was founded as the North American link in a world-wide network of stations. The British enterprise was directed by Edward Sabine and run primarily by the military under him. The involvement of the armed forces in such civilian work had increased sharply after the Napoleonic wars. By 1853 the Imperial government had decided to close down its operations but the Canadian Institute spearheaded a drive to maintain the observatory under the aegis of the provincial government. This section includes the Institute's memorial to the government concerning the observatory and a report of the government's assumption of responsibility for the operation.

In 1858-59 H.Y. Hind led an expedition to what is now Saskatchewan and Manitoba to collect a variety of information about the nature of the country. Hind's instructions to observers, included here, show the type of material his expedition wanted, and how they were to pursue their objective. Their observations comprehended cartography, geology, and meteorology. Meteorology was developed within an institutional framework in the 1860s, with a small network of observing stations reporting to the Toronto Observatory.

Accurate cartography and navigation require careful astronomical observations, and are necessary for the exploration and development of large and remote regions. Astronomy was therefore important in nineteenth-century Canada and was essentially utilitarian. The first government observatory in Canada was established at Quebec in 1850 under the direction of Lt. Edward Ashe, R.N., to provide time for the port of Quebec. The final extracts in this section deal with the observatory's foundation and function.

A. THE GEOLOGICAL SURVEY—SCIENTIFIC OR ECONOMIC ENTERPRISE?

Lord Durham, *Report on the Affairs of British North America* (London, 1839), pp. 12-13.

[British interests in North America] are indeed of great magnitude; and on the course which Your Majesty and Your Parliament may adopt, with respect to the North American Colonies, will depend the future destinies, not only of the million and a half of Your Majesty's subjects who at present inhabit those Provinces, but of that vast population which those ample and fertile territories are fit and destined hereafter to support. No portion of the American Continent possesses greater natural resources for the maintenance of large and flourishing communities. An almost boundless range of the richest soil still remains unsettled, and may be rendered available for the purposes of agriculture. The wealth of inexhaustible forests of the best timber in America, and of extensive regions of the most valuable minerals, have as yet been scarcely touched. Along the whole line of sea-coast, around each island, and in every river, are to be found the greatest and richest fisheries in the world. The best fuel and the most abundant water-power are available for the coarser manufactures, for which an easy and certain market will be found. Trade with other continents is favoured by the possession of a large number of safe and spacious harbours; long, deep and numerous rivers, and vast inland seas, supply the means of easy intercourse; and the structure of the country generally affords the utmost facility for every species of communication by land. Unbounded materials of agricultural, commercial and manufacturing industry are there: it depends upon the present decision of the Imperial Legislature to determine for whose benefit they are to be rendered available. The country which has founded and maintained these Colonies at a vast expense of blood and treasure, may justly expect its compensation in turning their unappropriated resources to the account of its own redundant population; they are the rightful patrimony of the English people, the ample appanage which God and Nature have set aside in the New World for those whose lot has assigned them but insufficient portions in the Old. Under wise and free institutions, these great advantages may yet be secured to Your Majesty's subjects; and a connexion secured by the link of kindred origin and mutual benefits may continue to bind to the British Empire the ample territories of its North American Provinces, and the large and flourishing population by which they will assuredly be filled.

The Canadian Review and Literary and Historical Journal, 2(1824), 377-95.

ON THE UTILITY AND DESIGN OF THE SCIENCE OF GEOLOGY, AND
THE BEST METHOD OF ACQUIRING A KNOWLEDGE OF IT;
WITH GEOLOGICAL SKETCHES OF CANADA

The study of Geology has of late years attracted the enthusiastic services of the first intellects of the age, by its novelty and usefulness; and by the grand and curious mechanism of the structure it attempts to explain. We know the Canadas to abound in valuable mineral products; and also in geological phenomena as interesting and instructive as they are neglected: we are therefore induced to intreat the attention of our readers to the results of such researches in extending national resources; and in advancing abstract science,—objects, in our estimation, equally honorable.

With this view, we shall briefly point out the importance and design of this branch of Natural History, and the best method of acquiring some knowledge of it;—concluding with a few sketches of remarkable localities in the Canadas.

It is only in appearance that Geology has been slow in engaging notice; for the philosophers of antiquity by no means withheld its fair proportion of their usual scholastic dreamings. It was natural, however, that its progress in modern times should be more tardy than that of Chemistry, Mechanics, or Pneumatics, &c. for they are based on the discoveries of the closet or the city, while the materials of the science now under consideration are gathered by the enterprising only, in distant and widely separated countries.

So great is the gratification of successful enquiry, that each department of nature will ever have its train of investigators; but geology, is not merely a recreation for the inquisitive; it exercises a prodigious and immediate influence on the civilization and prosperity of a people. It is gradually conferring on the operations of mining, (the true source of manufacturing greatness,) the same enlightened rules that chemistry has furnished to the economical Arts. It is banishing blind empiricism. Every day the ancient denomination of "Gentlemen Adventurers," assumed by the proprietors of Cornish mines, is becoming less applicable. It has collected, arranged, and examined, a great assemblage of facts, or rather of laws, and successfully applied them to the purposes of life. Certain invaluable substances, as magnetic iron ore, anthracite, coal, salt and gypsum, &c., have been shewn by it to exist in quantity, only in particular depositories —so that it is a vain waste of time and means to seek them elsewhere. The coal field of the north of England, has even been measured; and with the triumphant conclusion, that it will only be exhausted in 1500 years, at the present enormous rate of consumption. A few years ago, the miners of Derbyshire in England, threw all their white lead ore on the public roads, in ignorance of its nature. Very lately the Americans in building at Saguina in Lake Huron, were accustomed to fetch their limestone from Detroit, 130 miles distant, when it was plentiful in the bay adjacent. The officers of the Hudson's Bay Company, stationed at Fort William in Lake Superior, also have brought their limestone from Lake Huron, altho' it was to be procured 17 miles off, at the water's edge, near the base of Thunder Mountain. The early decay of the granite, of which Waterloo Bridge at London is built, is to be expected from the fact, which we have learnt from high authority, that the large crystals of feldspar, constituting so great a portion of the rock, is of the kind containing soda and therefore easily acted on by the weather. In an undertaking of so much moment, it is a matter of regret that the materials were not submitted to the judgment of a skilful geologist previous to their being used. . . .

Geology is the foundation of Physical Geography. On the nature of the rocks of any region depend its great features of mountains, vall[ey]s and plains, whose courses, dimensions and shape are derived from the position of the strata, and the peculiar outline, which each mineral mass, speaking generally, appropriates to itself. The same may be added of rivers, which are affected, also by the power of absorption possessed by their beds. Limestone being frequently cavernous, sometimes engulphs, partially or wholly, the streams flowing over it. Thus, part of the water of the Ottawa, immediately after making the descent of the very picturesque Falls of the Chaudière, enters a concealed chasm, and reappears in two places, the one in the middle of the river three fourths of a mile below, and the other as we are informed, about a couple of miles further down. Canada furnishes many examples of the characteristic features above alluded to. The shapeless, rounded massiveness of a granitic mountain is finely expressed by Cape Tourment, thirty miles below Quebec, which passed into the interior in huge flanks, now and then intersected by deep ravines of singular ruggedness and grandeur. Thunder mountain in Lake Superior presents a basaltic precipice 1400 feet high, of uncommon magnificence, faced by the usual rude colonnades. To these constantly recurring laws, often in beautiful groupings, we are indebted for the mouldering and fretted cliffs of sandstone on the St. Lawrence, a few miles above Brockville, and for those of limestones, at the Falls of Niagara, broken into stair-like ledges, overhung with large pointed tables of rock, and having their bases strewn with gigantic ruins. The pretty village of "The Forty" in Grimsby on Lake Ontario is close to a fine cliff of this kind. The Manitouline Islands of Lake Huron are full of them.

The botany of a district, as is well known to the student, and the agriculturalist is influenced essentially by its geology. Besides the operation of the latter on climate, the soil yielded by the disintegration of certain rocks is favorable to the growth of a particular order of plants, indifferent to another, and is often almost incapable of sustaining any kind of veg[e]tation. . . .

The extreme sterility of the countries immediately north of Lake Huron and Superior is owing to their granitic and other siliceous rocks; but much of the south shore of the latter Lake is held in irremediable barrenness by the vast quantities of sand and bowlders deposited there by the same great flood which poured abundance on the north coasts of Lakes Erie and Ontario in the fine calcareous clays which there prevail. We need scarcely add that the infinitely varied forms of animal life, their presence or absence in certain seas or countries, their number and perfection, are mainly produced by vegetation. Under these considerations, an acquaintance with the principles of geology appears to be indispensible to the general welfare. How extensive is the sphere of its controul. . . .

There are two views in which the prosecution of this science may be regarded; according as the student takes it up as an occasional amusement, or as the serious occupation of his life; designing, for instance, to illustrate the geology of his own country. Little labour will suffice to accomplish the first object: and truly fortunate is he who can occasionally escape from the collisions of commerce, or the strife of the passions, into the romantic scenery that surrounds our Canadian Cities;—to trace at every turn of the forest, in the curiously associated strata, their brilliant spars, and organic relics, the goodness and wisdom of the great Architect;—and his power in the convulsions and consequent devastation which the elements have at intervals caused. It is necessary that he should be acquainted with about an hundred rock masses and minerals, as granite, micaslate, basalt, quartz, serpentine, calcspar, &c. These he can never know from Books. Treatises on mineralogy are only useful to the advanced scholar;—to refresh his memory generally,—or to assist in the examination of unknown substances by their specific gravity, appearances under the blow-pipe, hardness, and cleavage, &c. &c. It seems almost impossible for the mind to embody and realise to itself a number of abstract qualities exhibited singly in books, and unaided, (as is the case,) by the approximation of the most important. A mineral held in the hand, presents to the senses a numerous group of leading characters. It is probable that a sufficiently comprehensive cabinet exists in most of the principal towns of U. and Lower Canada; to which, we feel assured, free access would be granted with particular pleasure. In case no such cabinet exist, from the fluctuation of society, common in colonies, Mr. Bakewell of London, (the author of many excellent works connected with these subjects,) is accustomed to furnish small ones at the moderate charge of £3 3s. Mr. Mawe in the

Strand, next door to Somerset house, sells collections, strictly mineralogical, (while those of Mr. Bakewell are geological,) for from 5 to 50 guineas. Both these gentlemen are in the habit of exporting to all parts of the world; so that a person resident in Canada, or in the East Indies, has only to send an order by letter, referring the party to an agent in town for payment, and he will find the package at his door in a few months. . . .

A correct and minute description of the geology of an extensive & complicated region is a task of no ordinary character; and especially on this side of the Atlantic. There are to be surmounted here, the difficulties incident to a new country, the greater portion of which is an unknown and unnamed wilderness, rendered impenetrable by displaced rocks, underwood and morasses, and therefore only to be examined in ravines and watercourses; in place of the cultivated hills and plains of Europe, illustrated by accurate maps, full of artificial sections by canals, mines, roads, wells, and quarries,—abounding in accommodation for the traveller, and what is still more essential, in fellow labourers, creating at every step, new light and new facilities. What a pleasing homage did science receive in the person of Le Duc, who during his geological travels through England, Flanders and Germany, on his arrival at any town or village was immediately claimed as the guest of the resident Prince or Nobleman, and was furnished likewise with the best local information, carriages, workmen, and intelligent guides.

In Canada, these researches on a large scale, become very expensive in hiring conveyances, by water and land to remote places: and the more distant these are from a dense population, the worse are the services and the more inordinate the demand. A government, or an associate body only, can afford to maintain a geologist in a distant and savage district like our upper Lakes from the great cost of the outfit. The necessary habits of extreme personal exertion from day dawn to dusk, contentment with coarse and often scanty fare, and the frequent exposure to cold and rains requires a powerful constitution; and the best is apt to fail under a continuation of these fatigues and privations. . . .

The country, generally is little aware of the value of such men as the owners of [mines and quarries]. Their capital, spirit and intelligence, are productive of manifold and most important advantages. New and extensive markets are opened for the produce of the distant settlers. Roads, mills, and stores are created, each individually a great benefit: Instruments of the first necessity in household affairs, and in husbandry, are offered, excellent in quality and at a cheap rate. But, perhaps the greatest blessing is, the example which these persons introduce—of a well-ordered family in the enjoyment of the comforts, proprieties, and accomplishments of superior life,—resulting from education and virtuous habits.

Canadian Agriculturist (1859), 235.

GEOLOGY AS A BRANCH OF GENERAL EDUCATION

Nor is it alone the miner, engineer, builder, farmer, landscape gardener, and painter that can turn to profitable account the deductions of geology. The capitalist who speculates in land, the emigrant, the traveller and voyager, the statistician and statesman may all derive assistance from the same source, and bring a knowledge of its facts to bear on the progress of their nations. So also the holiday tourist, the military officer stationed in distant countries, and others in similar situations, if possessed of the requisite knowledge, may do good service, not only to the cause of science, but to the furtherance of our industrial prosperity. Indeed we do not affirm too much when we assert that had one tithe of those who, during the last fifty years, have travelled or settled [throughout the Empire], been possessed even of a smattering of geology, these countries, as to their substantial wealth and social progress, would have been in a very different position at the present day. Their gold fields and coal fields, their mines of iron, copper, and other metals, take rank among the most important discoveries of the present age; and as the spirit of civilization is now evolved and directed, no progress can be made without those mechanical appliances to which the possession of coal and iron is indispensable, no facility of commercial intercourse without a sufficiency of gold, which has hitherto formed the most available medium of interchange. The assistance which geology has also conferred, and the new light its deductions have thrown on the other branches of natural science, are not among the least of its claims to general attention. The comparatively recent science of physical geography, in all that relates to the surface configuration of the globe—its climate and temperature, the distribution of plants and animals, and even touching the development of man himself as influenced by geographical position—can only lay claim to the character of a science when treated in connection with the fundamental doctrines of geology. So also in a great degree of botany and zoology; the reconstructing, as it were, of so many extinct genera and species has given a new significance to the science of life; and henceforth no view of the vegetable or animal kingdom can lay claim to a truly scientific character that does not embody the discoveries of the palaeontologist. In fact, so inseparably woven into one great system of life are fossil forms with those now existing that we cannot treat of the one without considering the other; and can never hope to arrive at a knowledge of creative law by any method which, however minute as regards the one, is not equally careful as concerns the other. Combining, therefore, its theoretical interest with its high practical value— the complexity and nicety of its problems, as an intellectual exercise, with the substantial wealth of its discoveries—the new light it throws on the duration of our planet and the wonderful variety of its past life, with the certainty it confers on our industrial researches and operations—geology becomes one of the most important of modern sciences, deserving the study of every cultivated mind, and the encouragement of every enlightened government.—

Advanced Text-Book of Geology, by D. PAGE, F.G.S.

B.J. Harrington, *Life of Sir William E. Logan, Kt.* (Montreal, 1883), pp. 137-8.

WILLIAM LOGAN TO SIR HENRY DE LA BECHE

24 April 1843

"From the fact that the Survey has been urged by the Legislature of the country, it is natural to infer that a great desire is felt by the enlightened part of the Canadian community to be made acquainted with the leading geological features of the Province; but the main object of the investigation is, no doubt, to determine the mineral riches of the colony, and it is not unlikely that a wish may be felt by its inhabitants to know the result or the probabilities of the survey long before it can possibly be completed."

Harrington, *Life of Sir William E. Logan, Kt.*, pp. 144-6.

LOGAN TO MR. RAWSON W. RAWSON,
SECRETARY TO THE GOVERNOR-GENERAL

Gaspé, 10th July, 1843

"My Dear Sir,—I have visited the Joggins, on the Bay of Fundy, and I never before saw such a magnificent section as is there displayed. The rocks along the coast are laid bare for thirty miles, and every stratum can be touched and examined in nearly the whole distance. A considerable portion has a high angle of inclination, and the geological thickness thus brought to view is very great. I measured and registered every bed occurring in a horizontal distance of ten miles, taking the angle of dip all the way along. Of course there has not yet been time to put together the facts thus collected; but when this is done, I shall be able to tell you every foot of what is in the crust of the earth in that part of it, for at least three miles deep. The whole deposit belongs to the Carboniferous era, and in one part of the section a multitude of coal-seams are exhibited. Mr. Lyell has stated them at nineteen, but they much exceed that. There is one thing, however, that Lyell has not mentioned, which is that the commercial value of this display does not by any means equal its geological beauty.

Of all the coal-seams exposed, I am sorry to say not more than two, or at most three, are sufficiently thick to be worked beneficially.

"In my examination of the neighbourhood of Bathurst, I saw only two coal-seams, but neither of them sufficiently thick to be profitably worked. One of them is six inches, and the other eight to ten inches.

"From all I hear, and something I see, it appears probable that the Carboniferous rocks do extend into Canada, but it is very problematical whether the Canadian part of the deposit will be productive. In this part of Canada there is a very favourable exposure of the rocks along the shores of the Gulf of St. Lawrence, and the various bays connected with it; and for the purpose of ascertaining the order of their superposition with accuracy, it is my intention to proceed around the coast with a canoe and an Indian to carry my instruments. Boating is too expensive, and not so independent a means of travel, and there are few roads of which to avail myself. I have with me, at my own charges, a young man of the name of Stevens, from Bathurst, a son of Mr. Stevens who established the Gloucester Mining Company, in New Brunswick. Knowing something of mineral exploration, having a dash of the necessary enthusiasm, and being accustomed to rough it in the woods, able to handle an axe, manage a canoe, and fit up a *camp*, as they call it, I anticipate with his assistance, and that of the Indian, getting along with economy and despatch. The nature and geological thickness of the formations that constitute the country once determined, the examination of their geographical distribution will be much facilitated. The chief difficulties connected with it will then be those of a physical nature in penetrating the woods.

"It is probable that to a geologist this part of Canada will present a great many more interesting features than the western division of the Province. It appears to differ considerably from what has been observed by the American geologists on the south side of the great lakes, in the State of New York. Disturbing forces have fractured the rocks, and thrown them into mountains and valleys. The country, therefore, abounds in picturesque scenery, in this respect far surpassing Western Canada; but for that very reason, in addition to its more northern latitude, it cannot be so fine a country for agricultural settlement. . . ."

Harrington, *Life of Sir William E. Logan, Kt.*, p. 151.

LOGAN'S JOURNAL

"*Tuesday, 18th. July.*—Two men came after me a considerable distance to-day, evidently watching my movements very narrowly. They spoke to me at last, and it seems they had considered me, from my various gambols about the rocks, out of my mind. Three clam-diggers did me the favour to inform me the same thing yesterday. I shall get much reputation here evidently."

Ibid., p. 167.

"Thursday, 14th. September.—... We have pitched our tent at Little River Cove, on the beach, and I believe all the inhabitants of the Cove have been to visit us, one after another. There are twenty-three families settled here. The number of fishing-boats is twenty-five. While occupied in examining the rocks on the other side of the brook, a multitude of the fishermen flocked around me, curious to know what I was about. One of them asked me if I was searching for buried money, and if the instruments I used indicated the proximity of hidden treasure. I explained the use of the instruments to him, and he seemed much gratified by it. . . ."

Harrington, *Life of Sir William E. Logan, Kt.*, p. 182.

LOGAN TO DE LA BECHE

20 April 1844

"I worked like a slave all summer on the Gulf of St. Lawrence, living the life of a savage, inhabiting an open tent, sleeping on the beach in a blanket sack with my feet to the fire, seldom taking my clothes off, eating salt pork and ship's biscuit, occasionally tormented by mosquitos. I dialled the whole of the coast surveyed, and counted my paces from morning to night for three months. My field-book is a curiosity."

Harrington, *Life of Sir William E. Logan, Kt.*, pp. 179-81.

LOGAN TO MURRAY

7 March 1844

"I have stated to the Governor that I should like to have some more defined arrangement as to the time the Survey is to be continued. He remarked that it had struck him that the voting of a certain sum without any knowledge as to whether that sum would be sufficient, was a very absurd sort of thing, and that it was but just that a more explicit understanding should be arrived at. I have spoken to Draper on the subject, and think he feels the propriety of putting the Survey on a firmer footing. . . . He has said, however, that unless it can in some way be indicated that value will be returned to the country for the expenditure, it is in vain to expect that the Legislature will support the Survey for the sake of science—in which opinion I thoroughly agree with him.

"In my interview with the Governor, I happened to say that before starting on my explorations in the spring I should, of course, make a report on the partial facts ascertained. He immediately replied that he would place it before the Legislature. . . .

"The object will be to produce an effect on the members. With the same view, I must get a house or a set of rooms for our collection. Managing this, we must put our economic specimens conspicuously forward; and it appears to me that in the exhibition of these, large masses will make a greater impression on the mind than small specimens. A sort of rule of three process seems to go on in the minds of the unlearned when they examine minerals in which they are interested. They are much addicted to judging of the value of the deposit by the bulk of the specimen shown.

"This induces me to say that I should like you to send to Montreal, as soon as it can be done by water communication in the spring, a thundering piece of gypsum. Let it be as white as possible, and put it in a strong box, similar to those I sent from below. If you come across the lithographic stone, let us have a huge slab of it, six or eight inches thick. . . ."

Harrington, *Life of Sir William E. Logan, Kt.*, pp. 229-36.

LOGAN TO DE LA BECHE

Montreal, 12th May, 1845

"My Dear De La Beche,—I have your despatch on Indian affairs, which renders it necessary that I should inform you of the position of my campaign in Canada.

"At the time I was appealed to to undertake the examination of the Province, a sum of £1,500 had been granted by the Legislature to defray the total probable expense. Of course I was aware that such an amount would be but a drop of what would be required to float me over twenty-five degrees of longitude and ten of latitude. But I undertook the survey, determined to work it out somehow or other, feeling the truth of the saying, 'Where there's a will there's a way.' Fearing that, if a term of years had been insisted on in the first instance, the Government might have been deterred from the undertaking, through an apprehension that they might not be able to get the consent of the Legislature, I said nothing about the matter. But upon being asked what my terms were, I named £500 sterling per annum for myself and £150 per annum for an assistant.

"These terms were considered very reasonable, and were granted. So at once I set to work. The £1,500 I thought might carry me on for two years, by which time I hoped to have had an opportunity to make friends, show the utility of the undertaking, and excite some interest in the subject among the legislators. I soon found that to make any impression it would be necessary to spend more money; that a business office, a museum, a chemist, and a laboratory would be required. At the end of the first year, accident threw in my way a young Pole, who had studied chemistry under Dumas at the Ecole Polytechnique, and brought good certificates of *capacité* from him. So I took the opportunity to urge the subject upon the Government. But a grand political rumpus had occurred. The French party had got out, and the English party had got in. Uncertain what they would be able to effect in the Legislature, they would give me no official reply, though they did not discourage me privately. On my own responsibility, therefore, I hired a house to serve for an office, museum and laboratory, at £120 per annum; provided chemicals and apparatus at my own expense, and arranged with the chemist for £200 per annum; and at the end of the second year I found that the Survey was about £800 in my debt.

"Avoiding politics as I would poison, I made friends on both sides of the question, and having induced our ministers to take up the matter, I got them to support it so far as £1,500 per annum, including all expenses for five years. Much to the surprise of my friend the Attorney-General, who took charge of the measure in the House of Assembly, there was not a dissentient voice upon the subject, the only subject of the session in which all agreed. Some of the members considered the sum too little, and £2,000 per annum was mentioned. So I was asked for an estimate of what would be required to place the Survey on an efficient footing. I made out that to pay my assistants at a rate adequate to a vigorous performance of their duties, and to do credit to the Government, on the basis on which I had put the Survey, it would require £1,950 per annum.

"An act was then passed voting £2,000 per annum for five years certain, for the employment of a suitable number of competent persons, whose duty it shall be, under the direction of the Governor-in-Council, to make an accurate and complete geological survey of the Province, and furnish a full and scientific description of the rocks, soils and minerals, which shall be accompanied with proper maps, diagrams, and drawings, together with a collection of specimens to illustrate the same; which maps, diagrams, drawings and specimens shall be deposited in some suitable place which the Governor-in-Council shall appoint, and shall serve as a Provincial collection. And duplicates of the same, after they have served the purposes of the survey, shall be deposited in such literary and educational institutions of the eastern and western divisions of the Province as by the same authority shall be deemed most advantageous. To defray the expenses of the survey and the arrears of expenditure already incurred, £2,000 per annum are applied for a term not exceeding five years, and I have to report annually in general terms.

"Such is my geological bill—now an act. I drew it up myself, and no changes were made in it. I should have liked very much to leave out the annual report, but I found it would not do, so I must be as cautious on that score as I can. I have recommended and obtained liberal salaries for my assistants. Murray gets £300 sterling per annum; so does my Pole. But, depend upon it, they shall do something for it. Murray works like a galley slave from the time he gets out of bed to the time he returns to it. De Rottermond has not done so much, but he has been in love, and is to get married on the 15th. inst. into a highly respectable family which has some French political influence. He is a young man, of gentlemanly manners, and, I think, of some energy, though he was completely knocked up in the forest last year, to which I carried him at his own earnest desire, just to show him there was no romance in the matter. . . . I fancy he will do, though, perhaps, he will require some management.

"Many parts of the country are so little known that Murray and I are in some places obliged to add topography to our geology. I wish I could let you see the map of our journey across from the St. Lawrence to Bay Chaleur. The length of our winding line is 111 miles, in which we dialled all the twists and turns of two rivers, one thirty-five miles and the other sixty-five miles, obtaining the bearings of the reaches by prismatic compass and the distances by Rochon's micrometer, and registering at the same time the quality, contents and attitude of every bed of rock we saw, with barometric heights, &c. The distance between the rivers we triangulated by means of well marked peaks, making it seventeen miles. I think you would say we deserve some credit for it. I have protracted the work on the scale of an inch to a mile. The distance in a straight line is seventy-five miles, and on comparing it with the same as determined by the latitudes and longitudes of its extremes on Bayfield's admirable hydrographical charts, we find that everything, without coaxing, falls into place. The bearings are identical, and there are only nineteen chains of difference between us in the distance. I have ordered three more Rochon's out, and I feel much indebted to Mr. Jordan for having mentioned the instrument to me. I think Jones[?] could get other orders for it from this country. If the economic facts of Canadian geology should turn out a negative quantity, the topographical facts may return some of the expense. I have made them available in getting the Survey continued.

"Now comes the application of all this egotism. Perhaps the Canadians are leaning on me for the Survey, and might think it not very handsome if I were to leave the country before the expiration of the five years. I am persuaded, though I say it, who should not say it, they will not find any one to take the trouble I do. It has been hinted to me that in continuing the Survey the Government have been in some degree influenced by the circumstance of finding a person who is a Canadian by birth considered competent to do the work. In the next place (but I do not feel that this weighs with me so much as perhaps it ought to do), I can get back my £800 only by saving on the £2,000 per annum for five years.

"But now comes another consideration, which perhaps weighs most. Just look at Arrowsmith's little map of British North America, dedicated to the Hudson's Bay Company, published in 1842. If you have not got a copy, send for one; the expense won't kill you, and there ought to be one in your Record Office. You will see that Canada comprises but a small part of it. Then examine the great rivers and lakes which water the interior between that American Baltic, Hudson's Bay, and the Pacific Ocean— some of the rivers as great as the St. Lawrence, and some of the lakes nearly as large as our Canadian internal seas, with a climate as I am informed, gradually improving as you go westward, and becoming delightful on the Pacific. It will become a great country hereafter. But who knows anything of its geology? Well, I have a sort of presentiment that I shall yet, if I live long enough, be employed by the British Government, under the Survey you direct, to examine as much of it as I can, and that I am here in Canada only learning my lesson, as it were, in preparation. How insignificant would be the expense to the British Government in comparison with the advantage that might result, and even the satisfaction of the enlightened curiosity would be cheaply purchased by what it might cost. I have been informed of coal in two parts of it—in the Saskatchewan territory, and in Oregon—in the Saskatchewan on the north branch of the river of that name, at Edmonton House, where it is burnt, and in Oregon near Fort Vancouver. But what the extent of the deposits may be, my informant (Sir George Simpson, the travelling Governor and General Inspector of the Hudson's Bay Company's establishment) was not able to say. They may be important. In Oregon, the value of coal for the supply of steamers protecting and connected with our new Chinese trade will perhaps soon be felt, and it might be an item worthy of the attention of the British Government in any settlement of the Oregon question with the Americans.

"When the British Government gave up the Michigan territory at the end of the last American war, with as little concern as if it had been so much bare granite, I dare say they were not aware that 12,000 square miles of a coal-field existed in the heart of it—larger than the largest in Britain, though the smallest of those belonging to the United States, which possess another of 55,000 square miles, and a third of 60,000 square miles. Saginaw Bay, on Lake Huron, cuts into the first, and Cleveland, on Lake Erie, is within thirty-six miles of the third, both ready to supply American steamers with fuel on the lakes, while ours on the same waters, in case of war, must depend on wood, or coal expensively transported from Nova Scotia or Cape Breton Island, or across the Atlantic from the United Kingdom.

"Taking all this into consideration, notwithstanding I have requested my brother Edmond, of Edinburgh, who has a friend in the East India direction, to make some inquiry into the matter, I fancy you will see that the chances are that I am tied to Canada. I feel grateful to you, however, for thinking of me, and the offer will do good. I shall not let the light of it lie hid under a bushel, but make it show my Canadian friends that geological investigations are something thought of in other parts of the world, and that if I do not accept pecuniary terms more advantageous than they give, it is because I am not influenced by mercenary motives in serving them.—Yours truly.

W. E. LOGAN."

Sir Charles Lyell, *Travels in North America* (New York, 1845), vol. II, pp. 193-4.

I never travelled in any country where my scientific pursuits seemed to be better understood, or were more zealously forwarded, than in Nova Scotia, although I went there almost without letters of introduction. At Truro, having occasion to go over a great deal of ground in different directions, on two successive days, I had employed two pair of horses, one in the morning, and the other in the afternoon. The postmaster, an entire stranger to me, declined to receive payment for them, although I pressed him to do so, saying that he heard I was exploring the country at my own expense, and he wished to contribute his share towards scientific investigations undertaken for the public good.

Canadian Journal, I (1853), 255.

'GOLD IN CANADA'

The extraordinary discoveries of Gold in California and Australia during the last four years, have so absorbed the attention of the public, that announcements, however important and advantageous, of the existence of other less dazzling but perhaps far more useful Mineral Deposits, have hitherto failed to excite that amount of public and private enterprize which, during other less Golden periods, would have stimulated to active exertion.

We shall not, probably, greatly err, if we venture to express the opinion that traces of a healthy reaction are now distinctly discernable in the Golden Fever of the day, lately so prevalent among classes in the enjoyment of permanent and remunerative industry.

The excessive toil and continued privation required on the part of the

Gold Digger,—not always with adequate results,—coupled with the well-ascertained fact, that those who continue to occupy themselves in the regular routine of established industry, more generally accumulate a sufficiency for independence and comfort, are happily arresting that unquiet spirit of adventure which has been so greatly aroused during late years.

We have elsewhere drawn attention to the admirable letter of Mr. Millett, on the Mineral Wealth of Nova Scotia. Coal, Iron, Copper, Barytes, and exquisite Marbles, constitute a noble Gold Field for our sister Province; and such treasures, with the exception of Coal, exist, too, in Canada East and West, independently of the more dazzling Metal to which we shall now call attention. Let us, however, suppose for a moment that a widely distributed auriferous soil, rivalling in richness the famed fields of Australia, were to be brought to light, and without due preparation and precaution on the part of the Provincial Government thrown open to the cupidity of those uneducated masses now crowding into the country. What effect would such a discovery have upon the construction of the vast system of Railways in progress or in contemplation throughout the Province? What difficulties would soon arise with our Gold worshipping and not over scrupulous or tractable neighbours? What a sudden and destructive check would the agricultural industry of the country receive, and all other branches dependant upon that expanding source of our present unexampled prosperity! What a flood of vice and crime would rush in to disturb, with its unhallowed and demoralizing influences, the quiet pursuit of intellectual and moral wealth, which now begins to display itself so vigorously amongst us! Here and there, throughout Western Canada, we find a painful solution of the question, in the case of a few misguiding or misguided individuals. Digging for Gold is a positive fact in various parts of Canada West. Delving sixty feet deep through the rich and teeming clays of the Valley of the Thames, and in the black bituminous shales below the veritable Golden Field (of Grain), the discovery of a few glittering lumps of Iron Pyrites is enough, in these days of Golden Fever, to turn men from remunerating Industry to waste their means in the hopeless search for Gold where no Gold exists. If digging for Gold under such unfavourable conditions is sufficient to secure the present ruin of a few, and to produce much local excitement, what might one expect if a rich auriferous soil in a thinly settled district were suddenly revealed to the eager and unfettered grasp of the uneducated labour of the country?

But does Gold really exist in Canada? Is it found in quantity likely to prove remunerative? To both of these questions we think we may answer in the affirmative with certainty. We may also hint to our Western friends, who are anxiously searching their own and their neighbours farms for the precious Metal, that the region which may truly be called Golden lies some hundred miles to the East and North-East of Western Canada. There appears no longer to exist any doubt that Gold is distributed over very considerable areas in Canada East, and in sufficient abundance to cause it to become a source of some anxiety to many interested in the progress of our Public Works and the general Industry of the Provinces.

We write, however, in the firm belief that precautionary measures are in progress, under the sanction of the Provincial Government, which will convert what would otherwise be a lamentable discovery, into a source of real advantage and profit to the country at large.

We have for some time past been aware of the existence of one powerful Association,—embracing some of the most distinguished individuals in the Provinces,—framed for the purpose of working a portion of the recently discovered Gold Fields on the Wage system, abjuring the Leasing system: a system at once ruinous to the labourer and destructive to order and morality.

We entertain and venture to express the opinion that whatever may be the extent of the Gold deposits in Eastern Canada, it is of the utmost importance that all Mining operations should be conducted systematically,— should be under Government supervision,—and that labour should not be dependent upon the success of individual exertions, but be in strict subordination to the Wage system.

Report of The Select Committee on The Geological Survey of Canada (Quebec, 1855), pp. 19-41.

W.E. LOGAN, ESQUIRE, OF MONTREAL, EXAMINED:—

How long is it since you commenced the Canadian geological survey?— *Ans.* I was applied to in the spring of 1842, by Lord Stanley, then Secretary of state for the Colonies, to know whether I would undertake a geological survey of the Province, and having agreed to do so, I came out and spent about four months in making a preliminary examination, in order to arrange a plan of work. Unfulfilled professional engagements required my presence in England, and I went back in December. For this preliminary work no charge was made to the Government; and the survey may therefore be said to have commenced when I returned to America on the 1st May 1843, or upwards of eleven years ago.

Can you give a short statement of what you have done up to this time?— *Ans.* It will be observed by a reference to the Reports of Geological Progress published, that the districts examined are as follows:

The Canadian coast and islands of Lake Superior, and two rivers on the north shore for distances of forty and sixty miles up. Here there has been shewn to exist an important copper region.

The Canadian coast and islands on the north shore of Lake Huron with

*distances of from twenty to seventy miles up four of its principal tribu-
taries.* Along the coast the copper-bearing rocks have been shewn to continue to some distance eastward of Lacloche.

*The coast of Lake Huron from the mouth of the Severn round by
Matchedash Bay and Cabots Head to Lake St. Clair; that of Lake Erie from
the vicinity of Chatham to the exit, and the upper part of Lake Ontario;
with most of the country included in the perimeter formed by these coasts
and a line from Toronto to Lake Simcoe.* In this have been shewn great
ranges of valuable building stone, of gypsum, and hydraulic and common
limestone, with extended areas of white and red brick clay, bog iron ore,
asphalt and mineral oil; while the structure, proved by the ascertained
distribution of the formations, shews that there can be no workable coal
beds in a part of the country, where even practised observers, without
due caution, would be liable in mistakes that might lead to great loss of
capital.

*The country in a general line between Lake Simcoe and Kingston along
the junction of the fossiliferous and unfossiliferous rocks;* in the former of
which are shewn the existence of a great range of valuable building stone,
as well as hydraulic and common limestone, with lithographic stone; and
in the latter enormous deposits of magnetic iron ore with whetstones,
plumbago, crystalline limestone and other materials; while the drift displays great areas of white and red brick clays, in some places covered by
extensive tracts of excellent peat and shell marl.

*The country between the St. Lawrence and the Ottawa, south of a
line from the vicinity of Kingston to Pembroke,* comprising a surface of
about 10,000 square miles, where in addition to great areas of peat and
shell marl, and clay fitted for common bricks and pottery, with bog iron
ore and ochre, great ranges of building stone, hydraulic and common lime
stone, and white sandstone fitted for the purpose of glass making, in the
fossiliferous rocks; and magnetic and specular ores of iron, lead ore,
some copper ore, plumbago, phosphate of lime, great and extensive beds
of crystalline limestone, sometimes giving good marble, barytes and traces
of corundum have been found in the unfossiliferous.

*The Ottawa from its mouth near Montreal to the head of Lake Temis-
camang, a distance of 400 miles, with many of its tributaries on the south
bank for distances of from twenty to forty miles up.* The economic materials in this are similar to those in the previous area and in equal
abundance.

*The north side of the St. Lawrence from Montreal to Cape Tourmente,
as far back as the junction of the fossiliferous and unfossiliferous rocks,*
comprising an area of 3000 square miles, in which have been found clay
fit for common bricks and pottery in great quantity, accessible in almost
every part; bog iron ore in large abundance, a profusion of iron and
manganese, ochres of various beautiful tints, tripoli or infusorial earth,

refractory sandstone admirably adapted for furnace hearths, white sandstone fit for glass-making, ranges of excellent building stone extending the whole distance, marble, and limestone fit for burning.

The south side of the St. Lawrence and the Eastern Townships from St. Regis to Etchemin River, a surface of about 15,000 square miles, a large portion of which is occupied by a mineral region of great importance, found to hold inexhaustible supplies of roofing slate and of beautifully variegated calcareous, and magnesian marbles, the latter resulting from a band of serpentine which been traced for 135 miles, soapstone in great abundance, dolomite, magnesite, chromic iron, whetstones, extensive intrusive masses of most beautiful granite, magnetic iron ore, occasional indications of silver-bearing lead ore, copper ore, and gold, while in the less mineralized part are good arenaceous and calcareous building stone, flagstone, white sandstone for glass-making, common brick and pottery clay, bog iron ore, peat, shell marl, and other materials.

The country between the Etchemin River, and Temiscouata portage road, in which many of the same materials as in the previous area will be found, but cannot yet be pointed out in a connected manner, the exploration having been only partial.

The coast of the Gaspé Peninsula from the Metis road by Cape Gaspé and Isle Percée to the mouth of the Matapedia River, a distance of about 800 miles, with several sections across the Peninsula from the St. Lawrence to Bay Chaleurs; the chief object of the exploration of this district was to determine the northern limit of the great eastern coal field of North America, spread out in the sister colonies; and as the carboniferous area lies unconformably on the inferior rocks, to ascertain whether any outlying patches might exist in the Peninsula. None such, however, have yet been discovered.

A large and valuable collection of specimens has been made to illustrate the economic materials, the minerals, rocks and fossils of the districts examined. This is preserved at the office of the Survey; and now that a suitable building has been placed by the Government at the disposal of the Survey, a commencement has been made to a classification and arrangement of the materials into two divisions, one to display the character and application of the useful materials, and the other the science of the whole subject.

The true bearing of geological facts, as parts of a whole, being unintelligible without the exhibition of their relative geographical positions, and so large a portion of Canada being still unsurveyed topographically, it has been necessary to measure accurately extensive lines of exploration and the maps resulting have proved of great value to the Crown Land Office. From this collateral work is derived a large part of what is known of the interior of the Gaspé Peninsula, where six streams have been measured; the Matane, the Chat, the St. Ann, the St. John, the Bonaventure and Great Cascapedia. It has shewn the courses of the Kamanistiquia and Michipicoten rivers on Lake Superior; of the Thessalon, the Missis-sague, the Spanish and the French rivers, on Lake Huron; in addition to 150 miles of the Ottawa and the whole length of the Mattawa. From it has resulted the improved delineation of the forms and distribution of a great chain of lakes in the rear of Kingston, and last year the course of the Muscoco from Lake Huron to its source; of the Petewawe from its source to its mouth; of the Bonnechere from its junction with the Ottawa to one of its sources; of the York branch of the Madawaska, with a sketch of the relations of various streams, from the tributary just mentioned to Balsam lake, the whole distance in these explorations and admeasurements being 500 miles.

Chemical analyses have been made of all the metallic ores, and such other useful minerals as required it, the number of which has been very great, and in addition of upwards of fifty valuable mineral springs, of a great collection of soils from both divisions of the Province, and of new mineral species....

What are the principal difficulties you have met with?—*Ans.* The principal difficulties I have encountered, independently of those unavoidably inci-dent to travelling in canoes up shallow rivers, and on foot through the forest, are those arising from the want of a good topographical map of the country. Accurate topography is the foundation of accurate geology. Unless you know the geographical position of every rock exposure that comes before you, you cannot tell the general relations of the whole, and you cannot make the physical structure of a district intelligible to yourself or to others. Without geographical position, the dip and strike of a rock are worth nothing, and the occurrence of a valuable mineral in two locali-ties distant from one another are just two isolated unrelated facts; while their topographical place being known, their dip and strike may imme-diately point to the probability, and guide to the search and discovery, of the same substance in a hundred places between. It thus becomes neces-sary in unsurveyed parts of the country to measure correctly, as I have already stated, long lines of exploration. But even in those parts which are settled, neighbouring townships having been surveyed separately and independently, and often not very correctly, it is next to impossible in putting them together to get them to fit. Lots, or portions of lots, that are in juxtaposition on the old maps given in to the Crown Land Office, are not so in the field; and in many of old surveys, lines in one and the same township, such as the township of Grenville for example, and others in the same neighbourhood, lines that on the paper are represented as straight, go staggering through the bush in zig-zags that would surprise an Indian hunter. In laying down work on such maps as these, it will be immediately seen, that if you have a useful mineral in two distant localities, such a mineral, for example, as serpentine, soapstone, slate, or such like, between which localities the observed structure of the country tells you the mineral mass should run in a straight line, and you should draw such a line from the one to the other on your paper, you might represent the mineral as occurring in lots where it was absent, and leave it out of those where it was present. Or supposing you followed the bed along its strike or direction from point to point, and then placed it on the lots in which it occurred, the result would be, that the course of your mineral would appear to have a multitude of what in this country are termed *jogs*. The geological inference to be drawn from the appearance of such on your paper would be, that the mineral band you were representing had been broken or dislocated by what are termed *faults*. The general bearing of your band would be incorrect, and might mislead you if you were depending on your result for further search; and if a map were published with these jogs, it would deceive geologists and mineral surveyors at a distance in respect to the general condition of the country's structure, making them think it was a disturbed one, and proper for the search of metalliferous veins, when it might have no such veins in it. Such a map would be more deceiving than one, on which the railroads should be laid down on the right lots in the old Crown Land plans of which I speak. No one could be deceived by the jogs in such a case, for the very nature and object of a railroad would proclaim to every one, that it could not have been so located unless the engineer had been insane. The incorrectness of some of the topographical plans, and the fact that we do not know which are right and which are wrong, makes it necessary for us even in surveyed parts, to count and register our paces over every road and line we go, taking the bearings by prismatic compass, and registering in its proper place every rock seen, with its dip and strike, and a short description of its character, and its economic and fossil contents, if it have any. . . . It will be easily understood, that this geographical work must unavoidably impede the rapidity of geological examination; and the necessity of so much measurement to fix the position of rock exposures, forces us, in order to make even a moderate progress, to examine fewer of them, or to give to each a shorter time than we would like, and thus, perhaps, to overlook some of its characteristics. . . .

What staff altogether would you think sufficient to put the survey upon the most efficient footing?—*Ans.* The present staff consists of five persons,—a directing and an assistant Geologist, a Chemist and Mineralogist, an Explorer and a Messenger. Taking all things into consideration, it appears to me that, to place the survey upon such an efficient footing as the Province could well afford, it would be necessary to add the two Topographers that have been mentioned, and the temporary aid they would require; to have an assistant in the museum, and the occasional services of an accountant, and two or three additional explorers, such as Mr. Richardson already mentioned with the temporary employment of miners to obtain by blasting specimens of economic materials, and of artizans to put them into useful forms. . . .

Do you think that you might derive much aid even from persons who are not strictly scientific men?—*Ans.* I am scarcely ever a day in the field in the settled parts of the country without getting a considerable amount of information from farmers and common labourers, particularly among such as are not haunted by the notion, that all our researches have the precious metals for their object. By a reference to the Report of Geological Progress presented to the Legislature this session, it will be perceived . . . how this immediately freezes up the fountains of communication. The settlers on the Ottawa, it appears to me, have got beyond the chance of such an epidemic, perhaps through the influence of some of the gentlemen I have named, and the *Ottawa Citizen*, which occasionally gives them a good sound geological leader. I have been informed, however, that when my friend Dr. Wilson first began his mineralogical researches, and used to carry home large blocks of stone to his premises, some of his neighbours imagined that, if he were not searching for gold, no other motive could reasonably justify his proceedings, and he might have suffered severely in parochial estimation, had not one, more sagacious than the rest, explained the matter to his own satisfaction, and that of the community, by announcing that of these stones the doctor made medicine. On the Ottawa more than any where else the settlers have appeared to appreciate what we were about, and have shown a readiness to give information and assistance. . . .

Would there have been any practical advantage if the [geological] maps had been sooner published, and the reports and other information more extensively distributed?—*Ans.* From the great demand there has always been for the annual reports, particularly by gentlemen from the United States, I am persuaded that a more extensive distribution would have led to a more early working of some of our economic materials. . . . Some time since, Mr. Samuel Keefer, c.e., called at the office of the survey, and requested permission to examine the map, on which we had represented the various formations of the Province; after poring over it some time he exclaimed, "Now I see where I am to get my materials;" on asking an explanation he informed me he meant the building materials required for the purposes of the railroad from Kingston to Toronto. By this I judge that if the map had been published sooner, it might have been of service to railroad engineers. Mr. Gzowski, in one of the reports of the St. Lawrence and Atlantic railroad company, gave public thanks to the survey for the information afforded him in respect to building materials. . . .

Can you mention any instances, which have come to your knowledge, of mistakes made in consequence of ignorance of the results you have already obtained, in consequence of their not being published at all, or insufficiently distributed?—*Ans.* On arriving at the seat of Government, then at Kingston, after my first examination of the Gaspé coast, and before any report could possibly have been published, I found that an Act had just been passed establishing the Gaspé Coal and Fishing Company. In conversation I expressed the opinion, that there was no coal where the company intended to sink for it. This reaching the ears of a gentleman interested in the adventure, he requested me to give him in writing, for the benefit of the company, the reasons for my opinion. The reasons resolved themselves into this, that the bituminous shales, in which the coal was expected, came almost visibly from beneath certain rocks, which had been ascertained by their fossils to be of more ancient date than the Carboniferous, and that the shales themselves holding *graptolites,* fossils never found so high in the series as the coal, it would be contrary to experience to find coal in them. The company which was an English one, having very probably submitted my letter to competent judges of the evidence, instead of paying a large sum of money down, (several thousands of pounds) for a property supposed to contain coal, which was the arrangement contemplated, got the conditions altered to the effect that they should pay down the money when the coal was found. Miners were subsequently sent out, but I have not heard of the discovery of the coal. The Act making this association a *Coal* Company was a mistake arising from our results at the time not having been published at all. . . .

Can you mention some cases of practical advantages resulting from the survey?—*Ans.* Limestone is almost an indispensable necessary of life. Those whose houses stand on the rock and who know it, never having experienced the want of it, can scarcely appreciate the inconvenience of those far removed from it. It has often given me great satisfaction to surprise a settler by shewing him, that he might have as much of it at his own door for six pence, as has for years in succession cost him, all circumstances taken into consideration, ten times the money, getting it from a distance. Almost every settler seems to know the calcareous character of the blue fossiliferous limestone, but very few that of the white crystalline rock of the Laurentian series. . . .

. . . The granular soils, from the crystalline limestones, are well known to be fruitful. On such a soil I have seen a field of oats every stalk of which was upwards of five, and a large number six feet high, with good grain at the head. The valleys underlaid by the rock have always constituted, in my mind, the main hope for the Laurentian country in an agricultural point of view; but the discovery of important ranges, largely composed of lime feldspars, greatly extends the prospect of advantage.

These rocks have been met with in several localities, from Abercrombie to the Sault-à-la-Puce in Chateau Richer; and as the Laurentian series in which they occur reaches from Labrador to Lake Huron, they are a subject of real importance to both sections of the Province. . . .

The practical advantages arising from the survey are only beginning to be felt. A great many substances have been pointed out that can be made available for domestic use, as the catalogue of them that has been published, and the specimens sent to the London Industrial Exhibition, very well shew. Discoveries usually lie dormant for some time after they are made. You may write about them, but cannot always get people to read, particularly if you circumscribe the distribution of what is written; and it often requires some accidental circumstance to bring them into operation. . . . I understand a branch railroad is nearly completed from [a slate] quarry to join the St. Lawrence and Atlantic at Richmond only a few miles distant. The moment it was observed by the farmers in the neighbourhood that here was a stone that might be turned into bread, every one began to search his own lot and try the schists that presented themselves. The result is that many find they have the same material and many more will continue to find it. An example is worth twenty reports. Few read but every one can see, and this example became for the vicinity, and in regard to this one material, what I wish to make the economic department of the museum for the whole public and all our materials. If a railroad should be carried up the Chaudière Valley the example will spread its effects there, as the slate will there be obtained in equal abundance. Several of the specimens of slate sent to the London Exhibition were from that valley. Soapstone is a material pointed out as existing in abundance. There are many establishments in the States whose business is devoted to the manufacture of it alone, and the Canadian localities are coming into operation. From what we have reported of peat and from the dearness of domestic fuel, a person in Montreal has commenced preparing and selling it for house use, at $5 per cord of 128 cubic feet unpressed, and $12½ for the same bulk pressed. He tells me that braziers and blacksmiths have been using some of it to their satisfaction, and I am aware that some enquiry has been making about it for the smelting of iron. It is used for such a purpose in France and other countries. It is known that 40,000 people are employed in France in the preparation of peat in various ways.

Do you think a fuller and more complete survey would produce similar results for the future?—*Ans.* There can be no doubt of it. In a geological survey of a new country, at first you obtain only a general sketch, as it were, of the subject, which you must fill up afterwards by degrees, and the more you enter into detail the greater will be your results. What you first point out will furnish the means of farther discoveries. The working of the first useful materials ascertained is almost certain to disclose facts that will point to the existence of others, and as you find different parts of the country fruitful in results, it may become advisable to extend researches in them. The very clearing of extensive tracts of forest, by producing a greater number of rock exposures, will occasionally render it expedient to go over them in greater detail than at first.

Can you give any instances of wasted capital for want of full geological knowledge?—*Ans.* The waste of capital in useless researches for coal in various countries is notorious. . . . Mr. Murray, in our first report, mentions a Canadian case,—"Many of the settlers," he says, "in the country underlaid by this formation," (the bituminous shales of the Hudson River group,) "seem to be strongly impressed with the opinion that it contains coal. In some instances I found them unwilling to listen to any reasons, which might interfere with their prepossession; and while a few, possessed of indications satisfactory to themselves, carefully conceal from general knowledge all information of the localities of their supposed buried treasure, through the apprehension, I was informed, that the Government would claim a right to all minerals discovered, others have proceeded more boldly to work, and have bored a considerable depth in search of the material. At Weston, on the Humber, I found that a company of adventurers had been partially formed, boring rods provided, an old miner employed, who, I believe, was a speculator in the concern, and the rock penetrated to a depth of 150 feet. Having, when two-thirds of the distance down, passed through a band of shale of darker color than usual, it was pronounced to be coal, and the work was continued in confident expectation of a large seam, until a deficiency of funds more than a want of hope caused the suspension of the operations." A bore hole was sunk at the upper end of St. Helen's Island in similar shales, with the same object. Another, I understand, was tried near St. Andrew's, in still lower rocks. It appears to me not improbable, that if the value of all the labor wasted in the Province in the mere transport of lime from great distances to spots, where it might have been procured with facility in the vicinity, could be computed with accuracy, it would amount to a much greater sum than has been devoted to the geological survey.

Can you give any illustration of the manner, in which a sound scientific basis leads to practical economical results?—*Ans.* A considerable portion of the science of geology is devoted to tracing out the distribution of the various formations, that come from beneath one another and spread over the surface of a country, the mode of representing these being by colors on a map. What is this, in an economic point of view, but a classification of its surface into parts, each of which will give useful materials peculiar to itself? So much is this the case that Dr. Buckland, in his Bridgewater

Treatise has shewn that a geological map of England is a map also of the distribution of its manufactures. Such a map will point out the limits to be observed in searching in new localities for materials that are known, and make every man's discovery of any useful material, not previously known, available to his neighbours in a hundred new places. For example, I was informed not an hour ago by Dr. Taché, that Mr. Gauvreau, of Quebec, has made from some of the strata on which the city stands a very good cement. It will immediately be seen by those acquainted with the geology of the country, that this is a discovery not for Quebec alone, but for hundreds of places between this and Missisquoi Bay, and for hundreds of places among the south side of the St. Lawrence below this. . . .

Copper ore has been pointed out and traced some distance in Inverness. The lode occupies a position in the Lower Silurian series of rocks (the top of the Hudson River group). In the report of progress for 1847-8, I stated that it was worthy of trial. Let us suppose that a successful mine were established on it. The result would give importance not merely to the ground in which the lode is immediately visible, but to all that belt of the Lower Silurian series in which the ore occurs, wherever the belt might run, provided the rock were in all cases in the same condition in respect to metamorphism; and the many cupriferous traces it contains in various parts, which otherwise would scarcely justify more than a passing notice, would become worthy of more serious attention. The general distribution of the rock could be made out from the map, and researches for the ore would thus be greatly facilitated.

Have you in your survey had as your principal object the establishment of new scientific facts, or has your attention been more directed to discovering and pointing out economic advantages?—*Ans.* The object of the survey is to ascertain the mineral resources of the Country, and this is kept steadily in view. Whatever new scientific facts have resulted from it, have come out in the course of what I conceive to be economic researches carried on in a scientific way. . . . The fact in Geological dynamics is a result of the examination that was necessary to ascertain the Northern limit of the coal field of New Brunswick, a most important economic investigation in so far as Canada is concerned. The double discordance discovered in the investigation rendered it possible, that outlying patches of the coal formation might lie unconformably on the Devonian, the Upper Silurian or the Lower Silurian of the Gaspé Peninsula; hence the propriety of the transverse explorations between the St. Lawrence and Bay Chaleur, of which two were completely across and several partially so. These have much narrowed the chance of coal there, but they have not quite exhausted it. They have, however, so much narrowed it, that it was considered proper to postpone further examination there, until other parts of the Country had received their share. The area of Canada is so large and the explorers so few, that we could not satisfy public expectation if we dwelt a very long time on one district.

The fact in Paleozoic Geography is a result of our economic researches up the Ottawa, combined with previous observations on the northern country by various persons. Our main object was to ascertain where the fossiliferous rocks began to dip northward, in order to determine what chance of coal there was in that direction. The Lower Silurian series was found wanting, hence the scientific inference. The probability of coal, however, if any exists in that direction, was out of the limits of the province. But though the coal was a negative quantity, other materials of an economic character were ascertained to exist.

The obliteration of the mineral distinction between vertebrate and invertebrate skeletons, results from economic researches for phosphate of lime, and the discovery may become of economic importance, not only to Canada but to other countries. In communicating it to the Director of the Geological Survey of Great Britain, I drew his attention to the fact that, if these phosphatic shells were found in any part of what are called the *Lingula beds* of his lowest series of fossiliferous rocks, in the same abundance that calcareous shells are in the calcareous rocks, the farmers of England would have to thank Canada for pointing out another source of this mineral manure....

The analyses of new mineral species, while they directly regard . . . scientific results, must always have also an economic bearing. You cannot tell whether a new substance is to be profitably available or not, until you have ascertained its properties? The analyses of mineral species led to our knowledge of the lime feldspars of so much agricultural importance to the Laurentian Country.

Thus economics lead to science and science to economics. The physical structure of the area examined is of course especially attended to, as it is by means of it that the range or distribution of useful materials, both discovered and to be discovered, can be made intelligible. A strict attention to fossils is essential in ascertaining the physical structure. I have been told that some persons, observing how carefully attentive I endeavour to be to this evidence of sequence, have ignorantly supposed the means to be the end, and while erroneously giving me credit as an authority upon fossils, have fancied economics to be sacrificed to them. In their fossil darkness they have mistaken my rush-light for a sun. I am not a naturalist. I do not describe fossils, but use them. They are geological friends who direct me in the way to what is valuable. If you wish information from a friend, it is not necessary that you go to him, impressed with the idea that he is a collection of bones peculiarly arranged, of muscles, arteries, nerves and skin, but you merely recognise his face, remember his name, and interrogate him to the necessary end. So it is with fossils. To get the necessary information from them you must be able to recognise their aspect, and in order to state your authority you must give their names. Some tell of Coal; they are cosmopolites; while some give local intelligence of Gypsum, or Salt or Building Stone, and so on. One of them whose family name is *Cythere*, but who is not yet specifically baptized, helped us last year to trace out upwards of fifty miles of hydraulic limestone.

My whole connexion with Geology is of a practical character. I am by profession a Miner and a Metallurgist and for many years, was one of the active managing partners in an establishment in Wales, where we annually melted 60,000 tons of copper ore, and excavated 60,000 tons of coal. It was my constant occupation to superintend and direct the minutest details of every branch of the business. A due regard to my own interests forced me into the practice of Geology, and it was more particularly to the economic bearings of the Science that my attention was devoted.

What do you think should be the object of a Provincial Survey in the economic point of view? And what kind of information should be expected from it?—*Ans.* The economic object of a Provincial Geological Survey in my opinion, should be to indicate in a comprehensive way, and in as short a time as possible, the natural resources of the Country, and the character and composition of its rocks, as leading to a knowledge of the origin and constitution of its soils. But the true mineral riches of a Country do not consist of its gold and its silver, but of those more common materials, which possessing little or no value as they lie in the earth, are yet capable of supporting a large amount of labor in receiving the forms, in which they become applicable to the wants of man. Where these are in fortunate combinations they may sometimes originate exports, but they can scarcely fail in any Country to be largely required for domestic use and afford employment to a great number of its inhabitants. Coal and iron are two of the materials, the presence or absence of which should be immediately ascertained. Limestone should be brought as near as possible to every man's door, and among the various substances capable of employing native industry, to which researches should be directed are common and refractory clays, building stones, slates, mineral manures, combustible materials, grinding materials, mineral paints, and a multitude of other things too numerous to be mentioned. Wherever these materials are found they are usually in quantities that are large and certain, and all that relates to them is easily calculated; and they would therefore constitute a safe foundation for Manufactures. Metalliferous veins are of course to be carefully attended to in a Geological Survey, but from their very nature, without considerable preliminary expense no approximation to true results can be formed. . . . In old mining countries analogies guide the miner to an opinion in new lodes in a known mineral district; but it would be dangerous to place too much confidence on these analogies, a thousand miles distant, in regard to results that in the nature of things are unavoidably attended with great uncertainty. All that can therefore safely be done in regard to mineral veins is to state their existence and describe their character where they are visible, leaving it to private enterprise to ascertain the expensive facts necessary to lead the way to a sound opinion in respect to quantitative returns. Even a moderately effective examination with this regard in any one locality would sweep away more than the whole funds at the disposal of the Survey for a season's exploration. It is not to be expected that the Geological survey of a country is to discover every economic mineral that exists in it. For centuries after the very best that could be made is finished, new materials will be brought to light; but one great duty of those who conduct such a survey, and perhaps it is the most essential, is to ascertain physical structure to the fullest extent the means placed at their disposal will allow, and to represent it. This is a work, the benefit of which will be felt for all future time, for as stated already, by this you arrive at a classification of the surface into parts, which, each in respect to a certain set of materials, limit the distribution both of what is known and what is to be known, facilitate discoveries, and make available to a multitude of his co-inhabitants, whatever mineral product the intelligence or good fortune of any and every individual may enable him to bring before the world. Even in an old country like the United Kingdom, where so great a number of her mineral products has been so long known, that we might despair of additions, it has but recently, it may be said, been thought wise by the people, that this classification should be accurately and minutely carried out, and new discoveries have been the consequence; its expediency therefore cannot be doubted in a new country at the beginning of its career, when there is so much to learn and so much to be made known respecting its mineral resources. . . .

Do you think it would be advisable to send a Geological Collection to the Paris Exhibition?—*Ans.* I certainly think it would be a very beneficial thing to send a Collection of Canadian Economic Minerals to Paris.

Do you think any great benefit resulted from our contributions to the London Exhibition?—*Ans.* In my opinion the London Exhibition was one of the most splendid and successful advertisements for Canada in regard to Minerals and every thing else that could have been contrived. She then became known to thousands upon thousands of just such people as she in various ways wants, who might otherwise never have dreamt of her resources.

Could you collect materials in the interval, which would do credit to the Province?—*Ans.* I could, with great exertion, make a Collection that

CANADIAN EXHIBIT, THE GREAT EXHIBITION, LONDON, 1851.

would very much resemble that sent to London, the masses, perhaps, not quite so large, and not so many pieces or specimens from each place.

Great Exhibition Catalogue (London, 1851), vol. II, p. 957.

BRITISH POSSESSIONS
IN AMERICA
Canada—New Brunswick—Nova Scotia
Newfoundland—Bermudas
West Indies—British Guiana—Falkland Islands

EIGHT dependencies of Great Britain are enumerated under this head. Of these, the most extensive collection of articles is that from the important possessions of this country in Canada. This collection, which is more particularly characterised below, is rich in raw materials and products. The other dependencies named are represented but by few exhibitors; but the articles exhibited deserve the attention of all interested in the commercial well-being of the countries and islands represented.—R.E.

CANADA

This vast and important territory is represented in the Exhibition by about two hundred and twenty exhibitors. The articles contributed by it are distributed among several Classes, but the raw materials preponderate; and of these a highly-instructive series is presented. The efforts which have been made by the Government at home to develop the mineral wealth of this colony have been amply rewarded by the success which has attended the explorers, and the results which in some measure are brought to notice in the Exhibition. A detailed account of the geological survey and its fruits will be found in this Catalogue. Many of the minerals exhibited must take an important commercial position on their locality and means of transport becoming known and developed. Among other and in reality more precious metals, the discovery of gold in the drift of the Eastern Townships along the southeast side of the Green Mountain range will be regarded with curiosity. Some fine specimens are exhibited, one of which weighs about a quarter of a pound. Copper promises to be more available for direct commercial purposes, and a cake of this metal is sent for exhibition. In this instance the ore has been smelted in Canada. A still more important mineral is the specular iron ore, of which a most valuable and important bed exists near the waters of the Ottawa, with abundant sources of water power, and ready means of transport. Most excellent iron is obtained from the bog-iron ore, wood charcoal being employed in its manufacture; it is comparable in its qualities with Swedish iron; and the stones and cast-iron works made from it are less liable to crack than those made in this country. . . . In addition to metalliferous minerals, . . . plumbago, asbestos, and lithographic stones, promise to become valuable sources of native wealth. Of these fine specimens are exhibited.

T. Sterry Hunt, Introduction to *Canada at the Universal Exhibition of 1855* (Toronto, 1856), pp. 415-17.

Canada has an area of about 40,000 square leagues; and the researches of Messrs. Logan and Murray, aided by those of Mr. Richardson, have already made known the geology of a great portion of this extent. According to the evidence given before a Committee of the Legislative Assembly, in October, 1854, it appears that the explorations up to that date, comprehended the shores of Lakes Superior and Huron, as well as all the great western basin of Canada, the valley of the St. Lawrence as far as the Gulf, the valleys of the Richelieu, Yamaska, St. Francis and Chaudière, that of the Ottawa and its branches as far as Lake Temiscaming, as well as almost all that part of Lower Canada south of the St. Lawrence, including the district of Gaspé. To these geological labours must be added the topographical surveys of several rivers tributary to Lakes Huron and Superior, of a great part of the Ottawa and its branches, as well as the surveys executed by Mr. Murray upon two lines of exploration between the Ottawa and Lake Huron, and the measurements of the principal rivers of Gaspé. All these topographical labours were only accessary to the Geological Survey, although necessary to its prosecution, and have greatly augmented the task of the Geological Commission.

The Annual Reports of the Geological Survey form at present about 1200 pages in 8vo., summaries of the geological researches of each year, with descriptions of the economic materials met with in the progress of the investigation, as well as researches upon the rocks, minerals and soils of the country, by Mr. [T.] Sterry Hunt, who has, since 1847, been attached to the Geological Commission in the capacity of Chemist and Mineralogist.

The inevitable expenses in a country where it has been necessary to carry on at the same time topographical and geological investigations, and to organize expeditions into regions still in a state of nature—have been such, that, notwithstanding the liberal sums accorded by the Provincial Government for these researches, it has not been without considerable personal sacrifice on the part of its director, that the Geological Survey has been carried on up to the present time. At the last Session of the Legislative Assembly there was accorded the sum of £2,000 for

the publication of a Geological Map of Canada, upon a scale of, $\frac{1}{600,000}$, (having thus a length of more than six feet by a breadth of three feet,) to be accompanied by a condensed summary of all the Reports which have yet appeared. It is proposed, during the continuation of the Survey, to publish each year, besides the annual Report of Progress, a *livraison* of ten plates of the characteristic fossils of the different formations of Canada accompanied by a descriptive text, and also to give geological sections, with a minutely detailed geological Map on a large scale, which will be published in several parts to appear successively. . . .

The Canadian government wishing to send to the Universal Exhibition at Paris a series of the economic minerals of the country, Mr. Logan was directed to collect them, and the minerals here exhibited, although in part, exhibited under the names of different individuals, were, with a few exceptions, collected by the personal care of the members of the Geological Commission. In order to indicate the geological relations of these materials, Mr. Logan has exhibited at the same time a map upon a scale of $\frac{1}{900,000}$, upon which he has brought together for the first time all the details of his geological labours; at the same time, as an explanation both of the map and the collection, we have thought proper to give in the little treatise which follows, a short account of the most interesting facts in the geology and mineralogy of Canada. We have added, moreover, a catalogue of the economic minerals of the country, and a small map, on a scale which is one-sixth of that about to be published. The geology of the neighbouring States is taken from the Maps of American Geologists, especially from that of Mr. James Hall.

For the geological facts, and for whatever relates to the physical structure of the country, all is due to Mr. Logan and his geological assistants; the mineralogy, as well as the chemistry of the metamorphic rocks and the mineral waters, is the result of the researches of Mr. T. Sterry Hunt, who has edited this little sketch.

Paris, August 1st, 1855.

The Canadian Statesman, Bowmanville, Thursday, Sept. 6, 1855.

CANADA AGAINST THE WORLD!

It must be interesting to every dweller in our portion of this great and glorious Continent, to hear such remarkable accounts of Canada at the great "Paris Exposition," and surely every man who has a heart or a head must feel that he owes a great debt of gratitude to Mr. Logan for his unceasing industry in the exercise of his decidedly superior practical genius, manifested in the study and development of every thing Canadian. Our country is rich in every thing that constitutes geographical greatness, but

SIR WILLIAM EDMOND LOGAN

were we without a Logan our greatness would never have been known to the old world.

Every Mail brings us fresh discoveries of his worth, and of the worth of Canada too, for though we are living in the Country itself, we would not have known half its worth, had we been without our Geologist. We have long entertained large thoughts concerning our adopted home, but we must confess that Mr. Logan and his contemporaries have presented things in a new light, his ability is almost universal, his effort indefatigable, and his whereabouts almost ubiquitary, always at hand when anything is wanted to forward the interests of Canada. We are so much enraptured with the love and pride of country, and we feel so much indebted to Mr. Logan for what he has done for it, that we are at a loss to express how deeply we feel on the one hand, or how much we are elated with patriotic pride on the other, certainly as Prince Napoleon said, Canada has much to be proud of, for the position she occupies in the world's fair at the Paris Exposition—and we cannot leave this subject without suggesting that no man in Canada deserves a Knighthood more than Mr. Logan, nor can our good and gracious Queen honor us more than by confering one on him, let him then be knighted, and let his coat of arms be henceforth the "Maple Leaf."

The Montreal Herald, 18th January 1856.

"WELL-MERITED DISTINCTION—Our contemporary, the *Gazette,* says that he has heard, 'from a reliable source,' that Her Majesty has been graciously pleased to announce her intention of conferring the dignity of Knighthood on our universally esteemed provincial geologist, [W.E.] Logan, Esq., in consideration as well of his current position in the world of science as of his labours at the London and Paris Exhibitions, in 1851 and 1855. *Honesta quam splendida!* How honourable are distinctions honourably obtained!"

The Ottawa *Citizen,* March 29, 1859.

THE GEOLOGICAL SURVEY

Of what use is it? To this question, a thousand answers may be given. No country undergoing the process of transformation from a state of nature to the cultivation of civilization, can have its resources properly developed, unless some measures be taken to ascertain what those resources are. The national wealth of the people inhabiting any particular region must be drawn from so much of the earth's surface as is contained within the boundaries of the country which belongs to that people. The savage races take no account of their mineral treasures, and avail themselves not of the agricultural capabilities of the soil. A semi-civilized community do not manage their affairs much better. The teeming riches of the earth remain unheeded, and the people are content to eke out a sluggish existence from the produce of the chase, or the flesh, milk, and skins of their flocks. A highly enlightened nation, such as that of the British Isles, France, or the United States, will turn everything possible into the means of their own subsistence, for home consumption, or by manual labour, into articles of commerce. An accurate inventory of all the materials found in the country is taken; scientific men are engaged to ascertain, by chemical analysis or otherwise, how far each substance may be made useful, or the object upon which industry may be profitably expended. One party is employed in seeking out the substances, ascertaining its position in the crust of the earth, the cost of procuring it, and the quantity that can be procured. Another, in the laboratory, ascertains its purity, value, properties, and capability of being made useful. With full information of this kind concerning the materials of economical value that may be procured, who will venture to say that a people are not better prepared to develop the resources of their country than another people ignorant of the treasures beneath their feet could be. As we understand the philosophy of the Geological Survey, this is its object. It is a national process, somewhat similar to what a merchant calls taking stock. Every merchant knows the value of this operation, and so should every nation. Canada has commenced taking stock, and the results of the Exhibitions at London and Paris bear witness to the benefits that have thus far resulted from the process. Our success at these recent displays of the material and manufacturing capabilities of nations, was, for the greater part, founded upon the information of the natural resources of the country, collected and methodically presented by Sir W.E. Logan. This information once procured is of use ever afterwards. Once it is ascertained what is our supply of the various metallic ores of iron, copper, lead, tin, silver or gold, or of rocks fit for lime, cements, building, roofing, flagging, polishing, or sharpening tools, whether of granite, gneiss, limestone, or sandstone; or for ornamental work or jewelry, such as the marbles, agates, carnelians, or malachites; or for manures, such as marls, coprolites, carbonates, or phosphate of lime; or what earths are fit for tiles, bricks, or pottery— when the agricultural geology, as it may be called, such as the distribution of the beds of sand, clay, and gravel in the different countries shall have been ascertained correctly, and the soils they furnish analysed—when all these, and a great deal more shall have been ascertained, and this knowledge transferred to the pages of a book, for the use of the public, then we shall be in a position to form an opinion of the value of the Province, what branches of industry will pay and should be encouraged—as well as what will be profitless and, therefore, not be pursued. A maxim of political economy is that a nation should confine its industry to the production of those things for which it is best adapted, and not those that cannot be produced unless at a cost beyond their value in the general market of the world. Without a Geological Survey, no Political Economist can direct the industry of Canada, or say what should or should not be done. For these, and a thousand other reasons that might be given, the Survey of the Province now in progress is of use. It is under the direction of a man of no ordinary ability. In all Sciences there are a few men who stand at the head of the list of Professors. In Geology there are only a few leaders. Such men as Sir R. Murchison, Sir Chas. Lyell, M. Barrande, &c. form a class composed of a limited number of members. In that class stands the director of our Geological Survey, Sir Wm. E. Logan. We hope, for the welfare of this country, that the Legislature will continue the Survey, feeling satisfied that the small sum of five or six thousand per annum out of the hundreds of thousands yielded by the revenue of this rich Province, will be returned a thousand fold.

... [B]ut what does it amount to? We can fancy some of our Legislative luminaries putting the question. To them the geological survey is a pastime—without results—or practical benefit, unworthy but of limited support, and should be placed in the category of other deserving schemes. We believe, nevertheless, that of all the investments of Government, none have been attended with such beneficial results as those which have followed the geological survey. . . . It has prevented mad speculations—restrained unprincipled schemers, directed healthy enterprise—and exposed our resources. What might the discovery of gold in the Chaudiere have done, had not its limit been defined? Half the Province might have gone wild with excitement. Agriculture might have been deserted, commerce abandoned, and the whole community in the desire to get suddenly rich, may have seen in the Chaudiere a second California—to meet only disappointment and wretchedness. But it became known, that gold could be extracted only by skilled labor—and that those engaged could only just get double wages. The same with coal. What sane man believes coal is to be found south of the Laurentian Range, in Canada, as it is explored and known? This fact established has saved the useless expenditure of hundreds of thousands of dollars. The black bituminous shales which throw out a flickering light on the application of heat, may easily deceive the superficial observer. But the perusal of Sir WILLIAM LOGAN's reports prove undoubtedly that we have no coal—that our rocks are of a prior age to the carboniferous era, and even men who packed coal on their farms, such as BOUCHARD and MENARD at Murray-Bay and the magistrate at Isle Perce, who would not hear that proof could be given that there was no coal—even such as these bow in submission to scientific truths. But the positive benefit of what the Province does contain also accrues. Slate may be taken as an example of this fact. A quarry of which has been opened at Kingsley, and is in operation, owing to mention of it being made in the geological reports. The same may be said of phosphate of lime—and the lime feldspars, and the hydraulic limestone. In short we know what is and what is not in the Province. And when was ever an erroneous fact attributed to Sir WM. LOGAN? Where have his statements misled—when has he failed satisfactorily to establish a practical truth, or a scientific theory? And what have we done for him in Canada? We have given him just as little encouragement as we could. For twelve years we allow him to toil on without praise or reward. And when he is absent in England on a mission which is to raise the credit and character of the Province—a result as successfully attained as positively as success can accrue—when this is being done, we have a nest of schemers trying to cut the ground from under his feet—to supplant him in his office, and to render the position which he has rendered so illustrious, subservient to a family interest. In Europe, where his fame is known—where his Canadian labors form a text-book—where the doctrines which he enunciates are welcomed as truths, he meets with accumulated honours and distinctions. Our true and staunch ally, the French Emperor, accords him the Legion of Honor—the Exhibition adjudges him the Gold Medal—the Geological Society presents him with the palladium WOLLASTON medal, the highest honor which a geologist receives, and HER MAJESTY distinguishes him by knighthood. What will Canada do? Will the cold welcome which political men give to men of science be continued? Will a few congratulations be the limit of recognition of a labor of thirteen years? We call upon the Government to adopt a different line of action. The report of the Committee of which Mr. LANGTON was Chairman, was presented last year, recommending the annual appropriation of £6,000. Such a grant should at once be made—but with this amendment. Sir WILLIAM LOGAN, with the disinterestedness which marks his career, stipulated that no recommendation should be made of any increased pay to himself. He asked merely that his assistants should be considered, and we see in the Report that he is still rated at £555 per annum. Such a man should not be paid one farthing under £1000 a year. A less sum would be a dishonor to a wealthy Province like Canada; and Sir WILLIAM's self denial should meet its reward. It is a known fact that tomorrow he could cross to the Republican Union, and as a mineral explorer, make his $20,000 a year; and that if he would, he could proceed to India, and obtain equal reputation to that which he has gained in Canada—and five-fold the stipend. Let it not be said then, that with us, merely the political adventurer meets his reward—that the only services which the State recognises are those rendered to party. Rather let Canada be looked upon as a country where distinction in science and art meet with generous encouragement, and where worth, when known, receives an ungrudging and cheerful acknowledgment. No child of Canada—and Sir WILLIAM LOGAN is an *enfant du sol*—has ever so distinguished himself. A debt of gratitude is due by the country—if payment is withheld the Province is disgraced.

Toronto *Leader*, 6 May 1864.

GEOLOGICAL SURVEY OF CANADA

Although we are destitute of precise information of the state of geological knowledge regarding Canada when Sir WILLIAM commenced his survey, what we do find is sufficiently suggestive. The fact really is that nothing was known of it. . . .

We may claim the result of the twenty years labors to be generally this, that for practical purposes we know, what mineral wealth we have, and what we have not. This information is no dubious, half credited belief. Like all results of well considered scientific operations, it has grown up as the information was disseminated, examined, weighed, and proved, to become a part of the national estimate of our resources. Mr. HALL, of New York, has borne testimony to the authority of the Reports of the Canadian Survey, not only with men of science, but with the more active class who seek information for the purpose of profitable investments: men who consider geology as the light by which they may safely operate in economic substances, and who need direction for the employment of capital. Canada has the satisfaction of knowing that no loss of capital, and waste of labor, with the misery consequent upon misdirected energy, has been a phase of provincial experience owing to the fruitless pursuits of what it was impossible to obtain. Mr. HALL gives his testimony that one million of dollars have been expended in abortive attempts to obtain "fossil fuel." Sir RODERICK MURCHISON computes that the money expended in England alone, before geology was understood would be sufficient to make a correct geological examination of the whole crust of the world. However, we have not been entirely without excitement on this matter. Six years ago, in June, 1858, Bowmanville coal was the latest thing, and no little excitement was kept up regarding it. One JOSEPH BALDSON discovered "a coal mine," on the property of Mr. BATES. A shaft was sunk for 60 or 65 feet, and borings were carried on 90 feet deeper. Enthusiastic public meetings were held. Everybody attended, and all made speeches who could, to have their fling at the "jargon of geology." The Mayor of Bowmanville came specially to Toronto to hold an interview with the Governor-General. A Vigilance Committee was appointed—and there was a howl of delight that all the scientific world was to be proved wrong, and "the uneducated, hard-working mining laborer" was to turn out the enlightened instructor of the age. Dr. CHAPMAN, to whom some specimens were submitted, easily recognized a detached cube of coal among some pieces of bituminous matter, and this piece of coal was declared to be found *in situ*. But the advocates of the discovery, no doubt in all honesty, but without reflection, endeavored to prove a little too much, when they declared that a quantity of coal had been taken from the "mine," and had been tested, and found excellent in the various blacksmiths' shops in the neighborhood. The quick eye of Dr. CHAPMAN at once saw the absurdity of the assertion, and the letters which he published had no little influence in exposing the imposture. Taking the data of the "miner," that the seam of coal was 6 ft. 5 in., and the bore hole was 3 inches, it was evident, that if the "coal" brought up had been in one solid mass it must have been less than the third of a cubic foot in bulk; therefore this actual coal, tested by the blacksmiths, must have been obtained elsewhere. But the fact is, as everybody acquainted with boring knows, that the matter brought up is a wet, pounded *detritus*, to which Dr. CHAPMAN gave the very emphatic word "slush," so that none of it could possibly have been consumed. A few days after, BALDSON, confessing his dishonesty, absconded. . . .

The days of Canada Coal mines may be said for ever to have passed away. Even at Quebec they have ceased to watch the shaft on Citadel Hill, the rude protection to which struck every visitor on his arrival. Perhaps all memory of it has passed away, and if we here chronicle these forgotten fables, it is because we remember that we are in a busy age of mining, and that there are dupes still to listen to the cheat. As the market is inundated with speculations, it would be well for the tempted to think twice before they embark their means, and some knowledge of the efforts made in the direction of "Canada Coal," may be no bad preparation to consider no few of the schemes put into the market. . . .

. . . To speak of the labors of Sir WILLIAM LOGAN is barely necessary. His reports annually published have constantly kept his name before the public. The two exhibitions at London and Paris, set upon his merits the seal of approbation from the highest and most renowned of every civilized country. In no metaphorical language he has positively brought honor on his native land. Living only for the work he took in hand, careless of honors, indifferent to amusement, with an ample private fortune to pass his life in literary and scientific ease, his hopes, his fears, his existence have been merged in this one work. At the end of five years after its commencement he was offered a salary of $10,000 to proceed to India. Perhaps in this quarter there was a field for a greater reputation than he was certain he would achieve here. Undoubtedly the road lay open to accumulated wealth and untarnished honors. In Canada there was nothing assured. Ignorant members of Parliament had assailed him, and an unsympathetic Executive thought they were extravagant devotees to science, by doling out their limited appropriation. But faith is strong and enthusiasm is untiring. Canada was LOGAN's home. Here he was born; here he had passed his youth, and his duties as it were seemed to him to lie written out before him. He had but to be patient, to abnegate, to be earnest, and his hour would come. And come it did. In time his character literally extorted from Parliament the consideration to which he was entitled. The conductors of this Journal look back with pride to the part they took nine years ago in enforcing the claims of the survey—for at that moment, there was a knot of men trying to cut the ground from under his feet and to appropriate his honors. The peculiarity of this illustrious man has been that everybody who has come within his influence has been carried away by personal devotion to him, and this feeling has told in the work he has to do. The very volume which leads to these remarks is proof of the fact. . . .

Henry White, *Gold: How and Where to Find It!* (Toronto, 1867), pp. iii-v.

The object of the following pages is, to supply the gold mining emigrant coming into Canada, as well as those already in this country, with the information necessary to enable them to conduct, with facility and certainty, the researches and operations which are essentially necessary to a successful issue in gold mining, as well as that of other operations of a similar nature. There can be no doubt but Canada affords greater accommodation, in consequence of its commercial facilities, for the successful prosecution of gold mining, than any other large gold producing country in the world, and there is less doubt, that a larger share of its mineral riches will ultimately fall to the lot of the intelligent and industrious mining student. The man who comes into this country, and makes himself acquainted (by the study of the following, or any other comprehensive intelligible work) with the conditions under which gold, or any other valuable metallic ore occurs, and is to be found, and the operations by which they may be most readily extracted from their native beds, will evidently find himself in advance and more successful than the less intelligent adventurer who trusts in chance alone for success.

   This work is therefore expressly prepared, with its accompanying geological maps, as a manual for explorers, and is designed to supply a want now very generally felt, respecting the Madoc Gold discoveries, and the occurrence of other valuable minerals in that (comparatively) unexplored and extensive region of Laurentian rocks.

   The work opens with a short description of the nature and character of the older rock composing the earth. Then follow in succession a further short description of the more recent rocky strata, their character, and distinguishing features, and metallic bearing veins, and shales, in the gold-

GOLD REGIONS OF CANADA.

GOLD:

HOW AND WHERE TO FIND IT!

THE

EXPLORER'S GUIDE AND MANUAL OF PRACTICAL AND INSTRUCTIVE DIRECTIONS

FOR

EXPLORERS AND MINERS

IN THE

GOLD REGIONS OF CANADA,

WITH

LUCID INSTRUCTIONS AND EXPLANATIONS AS TO THE ROCKY STRATA, PECULIAR SHALE ROCKS, VEINSTONE, ETC., IN WHICH

GOLD, AND MANY OTHER VALUABLE MINERALS,

ARE TO BE FOUND IN THAT REGION;

WITH

EASY MODES OF DETERMINATION AND ANALYSIS,

ACCOMPANIED BY TWO

COLORED GEOLOGICAL MAPS.

BY HENRY WHITE, P.L.S.,

Author of the "Geology, Oil Fields, and Minerals of Canada West,"
etc., etc., etc.

TORONTO:

PUBLISHED BY MACLEAR & CO.,

17 KING STREET WEST.;

1867.

bearing formation of this country, so that the explorer may readily distinguish the one from the other when he sees them, and search *only* in those places, stratas, and veinstones, which are pointed out in the body of the work, for his guidance, instead of making an indiscriminate and hopeless random search in all rocky stratas he may meet with.

There is also given such a description of the appearance, nature and characteristics of the Laurentian or gold-bearing rocks, as will enable the explorer at once to distinguish them from the rocks of the Silurian formation lying on its south side. Besides which he is also refered to the accompanying geological maps, on which he will see the line of contact between the two formations, distinctly marked, so that he need not in any case mistake his geological or geographical position in relation to the gold-bearing formation.

A description of all the most valuable minerals that are known to exist in, and belonging to that formation, is also given in detail, with easy modes of determination and analysis, and much valuable information, respecting the proper places and veinstones in which to search for gold, &c., &c., with plain and practical methods of ascertaining its existence in Talcoso or chloritic shale, quartz rock, iron pyrites, alluvial deposits, red ochre, or black sand.

The author has endeavoured to keep as clear as possible from technical terms and incomprehensible phraseology, thereby rendering the work as plain, practical and intelligible as possible, so that it may be easily understood by every person who may read it.

The want of a plain, cheap, and comprehensive work of this kind is severely felt in Canada just now, and the great tide of explorers and gold seeking emigrants that will, on the opening of navigation, visit the Gold Regions of Western Canada, will make it doubly so.

The few works we have in Canada are either too elaborate, too scientific, and too expensive, or are the production of foreign countries, and adapted only to their geological conditions and mineral characters, and wholly inadequate to the wants of the Canadian explorer.

To supply this want in the requirements of our young, but great and undeveloped mineral country, and to place in the hands of the venturous explorer, such information and guidance as he must necessarily require to be successful, and, without which his labour will be in vain, is the object of this book, and the author reasonably believing that his long professional practise as a P.L.S. and mineral explorer, through the rocky wilds of Canada, render him not unfit for the compilation and accomplishment of the following exposition of the undeveloped mineral resources of the Laurentian formation of Canada, of which the following pages principally treat.

TORONTO, April, 1867.

B. SCIENCE, THE MILITARY, AND EMPIRE. MAGNETIC, METEOROLOGICAL, AND ASTRONOMICAL SURVEYS

Great Britain, Meteorological Office Archives (Bracknell, England), MS. 78, pp. 71-2. Reprinted by permission.

EDWARD SABINE TO GENERAL SIR JOHN HARVEY   Woolwich, 22 July 1844

Dear Sir John:

You are aware no doubt that in the last few years magnetical observatories have been established by our own and other governments in different parts of the globe, for the purpose of determining by systematic and simultaneous observation certain fundamental facts in the magnetical and meteorological sciences. Our own Country has taken the lead in this undertaking, by prescribing the system, and the times of simultaneous observation: we have also occupied stations, as at Toronto, St. Helena, the Cape of Good Hope, Van Diemen's Land, Simla in the Himalayas, Madras, Bombay and Singapore, so spread over the globe as to bring the phenomena taking place at the same instant over its whole surface into immediate comparison.

The results of this extensively combined operation have already been extremely surprising and curious; they have shown the universality of certain magnetical affections or impulses which the globe is continually receiving; but of which we know neither the laws nor the origin. We appear, however, to have some prospect of acquiring a knowledge of both, by studying the difference in the effect which the impulsive force acting on the whole globe produces in different parts of it. For example, the effect shewn at stations however widely distributed over Europe is nearly identical; the effect at all the stations yet established in the North American Continent is also identical; the American stations agree with each other; as do the European ones; but striking dissimilarities are found, when the two continental systems are compared with each other; and it is very conceivable that these dissimilarities may afford a clue to the cause of the whole phenomena. You will perceive at once that it must be extremely important in this point of view to know what passes at an intermediate station, such as Newfoundland, where one system must be in progress of transition into the other.

As our Government have employed the officers and N.C. officers of the Artillery to make the observations at the station to which I have referred, the Regiment in general has naturally taken an interest in the enquiry; and the officers of the Company which has just embarked for your Government have shown a disposition to take part in it as far as may be in their power. Well knowing your public spirit, and readiness to promote whatever may contribute to the advancement of knowledge or to the credit of those who have a desire to employ their leisure time usefully, I have not hesitated to supply them with the instruments which are used in our observations: to enable them under your sanction and encouragement, to take such share in the general system of observation as may be found to consist with their duties, and with circumstances.

I have placed the instruments in the charge of Lieut. Brethingham, who has taken such pains to profit by the instruction, which has been given him here, where we have a sort of headquarter establishment from Lieut. Riddell, who was himself the officer in charge of the Toronto Observatory. The instruments are complete for all the purposes of a magnetic and meteorological observatory, except in wanting a chronometer and a theodolite, both articles I have supposed may be obtained temporarily, from the Surveyor General's Office in the Colony. No special building is necessary to be erected for their use, as an apartment in the ground floor, detached from other occupied buildings, will perfectly answer the purpose.

I hope to send you in two or three months the first published volume (now in the press) of the observations made in the British Colonial Observatories, by which you will be enabled more fully to apprehend the general scope and object of the undertaking. . . .

To His Excellency Major General Sir John Harvey
Newfoundland

MO/Hist. Letters (29)/13E

RIDDELL TO SABINE   [Toronto] 5/3/1840

I hope you will fulfill your promise of coming out here this year or the next. I have not found anyone here who appears to take the least interest in the operations or on whom I could rely for the least assistance. There is a College and three or four Professors but I have only seen one of them the shallowest little wretch I ever met—my only chance is in the arrival of some officer who may have a turn to science—and I am not likely to have everything at the observatory in a sufficiently satisfactory state to allow of my leaving it at all before the close of the year—otherwise I should have liked a fortnight trip up the Lakes very much—with the Force needles I should have all that is required for the dip and intensity but should have no means of ascertaining the variation which I suppose an equally important claimant as either of the others or at least the dip. If you are unable to come out and make a magnetic tour it will, I should think, be very desirable that I should make as many as I can in different parts. There is such an immense chain of steam communication on both sides of Toronto that in two trips of a fortnight each or less I might make observations at places more than 1000 miles apart.

Canadian Journal, 1(1853), 145.

MEMORIAL OF THE CANADIAN INSTITUTE TO THE THREE BRANCHES
OF THE LEGISLATURE TO CONTINUE THE ROYAL MAGNETIC
OBSERVATORY UNDER PROVINCIAL MANAGEMENT

To the Honorable the Legislative Council of the Province of Canada, in Parliament Assembled.
The Memorial of the undersigned members of the Canadian Institute, Humbly Sheweth,—

That your Memorialists have heard with much regret that Her Majesty's Government has determined to withdraw the Detachment of the Royal Artillery at present employed in making Magnetical and Meteorological Observations at the Observatory at Toronto, and to maintain that establishment no longer.

That your Memorialists being members of a Society incorporated by Royal Charter, for the purpose of promoting the cultivation of scientific pursuits in Upper Canada, view with great concern, the discontinuance of the only observations made systematically and upon a large scale, on any class of natural phenomena, in British North America.

That as regards the science of Terrestrial Magnetism, your Memorialists believe that all which has yet been effected in that subject, has but opened the way, to wider and more general enquiries; that the period over which the observations at present extend, is much too short to have elucidated completely the various annual and secular changes which it has brought to light, and that a prolongation of those researches, more particularly, which have indicated a connection subsisting between the magnetic variations and the *solar spots*, and a secular period in both variations, is eminently recommended by their novelty and interest.

That your Memorialists believe that the discontinuance of the observations so long and so systematically made in every department of Meteorology at this establishment, will not only deprive all those interested in that difficult and intricate subject, of a centre of reference, of comparison, and of support, the local and immediate value of which is, perhaps, more generally felt, than that of any other class of observations, but will also cut off the possibility of a large class of highly important enquiries, more particularly those which relate to the gradual change of climate which Canada is supposed to be undergoing, to their influence upon Agriculture, and to the periodical recurrence of seasons marked by peculiar manifestations of disease, and other important practical characteristics; which require a long, unbroken, and strictly comparable series of observations for their solution.

That your Memorialists conceive that it will be a reproach to a country so populous as Canada, of so large a public revenue, and possessing a University so largely endowed, if it suffers an establishment to fall to the ground which is of confessed scientific importance, and in whose continuance scientific men in the United States and elsewhere have repeatedly expressed their warmest interest.

That your Memorialists believe that the time has rather come when its operations should be placed upon a less restricted basis, and be extended from the special objects for which it was originally founded; to make it a centre of reference for all that large class of pursuits which involve periodical phenomena; and to include those higher departments of science, and more particularly of Astronomy, to which every Canadian must aspire to see his country one day contribute.

Your Memorialists, therefore, pray that your Honourable Council, will be pleased to take such steps, as to your wisdom may seem best, to effect the further continuance, by Provincial authority, of the Observatory heretofore conducted at the expense of the Imperial Government in Canada, after the withdrawal of the Military detachment; by placing it in connection with the Provincial University, or by maintaining it as an independent Provincial Establishment.

And your Memorialists, as in duty bound, will ever pray.

The following extracts from a correspondence printed by the Royal Society for the information of its members, in 1850, are interesting, in connection with the subject of the foregoing petition, and well calculated to assure the public that in placing the Observatory at Toronto upon a stable basis, the Government will only be carrying into effect what has been called for by men of the most eminent science in England and the United States. A country, whose public revenue approaches a million pounds currency, (£842,184, in 1851,) and whose enormous and costly public works attest at once the vigour of its resources, and the boldness with which it can be applied in measures of national importance, cannot be excused from bearing also a modest share in those burdens,—if they can be so called,—which a wise recognition of the claims of science has added, in almost every civilized land, to the necessary cost of civil administration or material development. For what, after all, is Science? It is nothing but the investigation of those laws of nature and properties of matter, our acquaintance with which is the foundation of all national prosperity; and which, once mastered, enable us to subject the one, and bind the other, to our car of triumph. No country, capable of reciprocating the advantages she derives from others in this respect, can justly refrain from doing so.

Canadian Journal, 1(1853), 282-3.

THE OBSERVATORY

In our last issue we informed our readers that the Magnetic Observatory at Toronto, established by the Imperial Government and supported by them for a period of twelve years, had been taken in charge by the Provincial authorities, with the intention of being retained as a permanent establishment: we are now able to give more detailed information on the subject.

Some time in February last, **Captain Lefroy** received orders from the home-government to pack up the instruments, dismantle the observatory and return home with the military detachment which had been, under his superintendence, employed in the observations. With his usual zeal and energy, he lost no time in bringing the matter to the notice of his Excellency the Governor General, urging the importance and interest of the scientific results that might be expected from retaining an observatory complete in all points and which had already earned a reputation second to none throughout the world. In these representations he was powerfully backed by the petitions of our own and kindred societies in both sections of the Province. With most praiseworthy promptitude and liberality, the Provincial authorities at once communicated with the Imperial Government offering to purchase the equipment of the observatory in full, and in the same spirit they were responded to, and the negociation completed without delay. The munificent sum of £2,000 voted for this purpose in the last session of Parliament gives a striking and most pleasing proof of the esteem in which Science is held in this country.

In the meanwhile Captain Lefroy had returned to England, leaving, however, the Military Detachment behind, and formally placing the Observatory, according to his instructions, under the charge of Mr. Cherriman. The Magnetical Observations had been in part interrupted by the introduction of Iron during the process of packing some of the Instruments which could not be left behind, and also by nearly all the Instruments having been dismounted for the purpose of final verification. Their adjustment of course occupied some time, but it is now completed, and the full observations are now made as before. The Meteorological observations have never been at all interrupted. Instruments to replace those taken away, besides others which it has been thought advisable to introduce, have been ordered from England, and are daily expected, and certain necessary repairs and alterations will be commenced as soon as the plans for them can be procured.

The Military Detachment so long employed on this service, has been permitted by Her Majesty's Government to remain here for so long a period as may be necessary to enable Mr. Cherriman to make a report to

SECOND OBSERVATORY, TORONTO.

His Excellency, of the staff that will be required, and of the steps that may be advisable to render the establishment permanently effective and complete.

We cannot conclude without congratulating the Province upon the completion of arrangements which secure to Western Canada this extensive and well appointed Magnetic and Meteorological Observatory, under a gentleman whose distinguished career at the University of Cambridge is sufficient guarantee that all the interests of Science will be as industriously and efficiently maintained as they have hitherto been, within the same walls, under that management which has given to it the wide spread and exalted reputation it now enjoys throughout the scientific world.

Canadian Journal, 3(1858), 99-103.

CANADIAN INSTITUTE, PRESIDENTIAL ADDRESS

Turning once more to domestic matters, I shall beg permission to occupy your attention with a few observations touching the Magnetic and Meteorological Observatory in this city; which I am the more induced to do from a doubt whether the intrinsic value of the establishment, and its effects in making Toronto known throughout the civilised world as the seat of this Observatory, are sufficiently valued and appreciated among us.

Established at the instance of the Royal Society by the Imperial Government, this Observatory formed one of a chain of stations which were, almost simultaneously, called into existence, either by national support or private liberality, over the whole face of the globe, and were designed, in connection with exploring expeditions, both by sea and land, to furnish the data by which it was hoped the secrets of that mysterious agency, the earth's magnetic force, might be laid bare. Its existence was prolonged much beyond the period which had originally been proposed, and which was found quite insufficient for the accomplishment of the work,—the liberality of the Imperial Government being successfully appealed to by the same learned body to whose exertions its institution was due,—and when ultimately the period arrived when its abandonment was no longer to be deferred, the offer was made to transfer it to the Province with its complete equipment, free (with slight exceptions) of cost, and subject only to the condition of its permanent maintenance. It is a just ground for congratulation that this generous offer was accepted, and that the Province has responded to the call of Science, not only by providing an ample endowment for the Observatory, but by replacing the temporary wooden structure in which its operations were formerly conducted, by handsome and substantial erections of stone. It may fairly be allowed to the members of this Institute to indulge the belief that these desirable results were effected, in part at least, by the urgent representations which they and their President at the time made to the Government; nor will the pardonable pride they may feel in the matter be lessened by the knowledge that, out of all the Colonial Observatories which were in similar circumstances, this is the only one the retention of which has been accomplished. The outlay on the Observatory for its erection and equipment from first to last has probably exceeded £5,000, and I believe that in completeness and efficiency it is not surpassed, if even equalled, by any observatories in the world. Three large quarto volumes containing the observations made here, have already been published by the Imperial authorities (and a fourth is yet due), carrying the name of Toronto into all parts of the earth where science is cultivated; and so remarkable and valuable have been the theoretical results deduced from them (to which I shall presently more particularly allude,) that it is not too much to say that the name of a Canadian city, which will be sought for in vain on maps twenty years old, has now become, by means of its Observatory, familiar in the mouths of European savans as a "household word."

Very few, if any, subjects of inquiry are of greater interest and probable importance to science, than that of terrestrial magnetism. Practically familiar, as we have been, for a long course of years, with many of its phenomena, the theories invented to account for and to explain them were more owing, as has been well remarked, "to the boldness of ignorance than to the just confidence of knowledge;" and the "want of a foundation whereon the advancement of that science, on inductive principles, might be based, was strongly and extensively felt."

The objects of the Magnetic Observatories were, as I understand, to investigate the periodical variations in the terrestrial magnetic force, by suitable instruments and methods; to separate each from the others, and to seek its period, its epochs of maximum and minimum, the laws of its progression, and its mean numerical value and amount; that, by a combination of the results attained, a general theory of each, at least of the principal periodical variations, might be derived; and tests be thus supplied, whereby the truth of physical theories propounded for their explanation might be examined. With the observation of their *periodical variations*, was combined a comparison with meteorological variations of a periodical character; which together with those *"secular changes*, which with slow but systematic progression alter the whole aspect of the magnetic phenomena on the surface of the globe, from one century to the next, and which in their nature are not improbably intimately connected with the causes of the magnetism of the globe itself," were deemed subjects of inquiry of the highest importance by "those who, by the inductive process, would seek to ascend to general laws and to the discovery of physical causes."

It is beyond my province, and still more beyond my power, to attempt to trace and define the progress of these observations, and the results which, so far, have been attained. But I am justified in remarking, that the observations recorded here in Toronto, occupy a very high place in the estimation of those scientific men whose attention is devoted to this interesting branch of science. Major General Sabine, himself a member of the Committee of the British Association for the Advancement of Science, by which the attention of Her Majesty's Government was solicited to the expediency of establishing fixed Observatories in the British Colonies, has remarked that the observations at the station at Toronto considerably exceeded 100,000 in number: that "Toronto is the first and, [as] yet, the only station at which the numerical values at every lunar hour of the lunar-diurnal variations of the three elements," viz.: the horizontal direction, the dip, and the intensity of the magnetic force, "have been published." And he pays this handsome tribute to those who have had charge of this Observatory: "It is with much satisfaction, and with a well-deserved recognition of the pains which have been bestowed by the successive Directors of the Toronto Observatory and their assistants, that I am able to refer to the determinations of the absolute values and secular changes of the three elements contained in the third volume of the Toronto Observatory, in evidence that the instrumental means that were devised, and the methods which have been adopted, have proved, under all the disadvantages of a first essay, sufficient to determine the data with a precision which is greatly in advance of preceding experience, and, as far as may be judged, equal to the present requirements of theoretical investigation. This is the more deserving of notice, because Toronto is a station where the casual and periodical variations, which it was apprehended would seriously interfere with the determination of absolute values, are unusually large. We may derive, therefore, from the results thus attained, the greatest encouragement to persevere in a line of research which is no longer one of doubtful experiment, and to give it that further extension which the interests of science require."

That the task of determining the true laws of the phenomena observed, is, as yet, very far from being accomplished, cannot be denied; but this should not for an instant create doubt or hesitation. Nearly two centuries have been found insufficient to work out all the consequences of the principle of gravitation. The discoveries, with regard to magnetism, are apparently only opening out to view wider and wider fields of inquiry. Professor Faraday, in speaking of the coincidence which has been observed between the maxima and minima of the daily magnetic variation in declination, and the increase and decrease of the solar spots, remarks that "the observation of such a coincidence ought to urge us more than ever into an earnest and vigorous investigation of the true and intimate nature of magnetism, by means of which we now have hopes of touching, in a new direction, not merely this remarkable force of the earth, but even the

like powers of the sun itself." To this it may be added that a similar antici-
pation may be indulged with regard also to that luminary which "governs
the night," when we remember that remarkable discovery of the variation
in the earth's magnetic force, which has been shewn by General Sabine—
chiefly from the Toronto observations,—to depend on the place of the
Moon.

In addition to the foregoing testimony in reference to the Observatory,
Major General Sabine, at the Dublin meeting of the British Association,
last year, instituted a comparison between the observations at Toronto
and those made by Captain McGuire and the officers of H.M.S. *Plover*, at
Point Barrow, in 1852-3-4, when employed in searching after Sir John
Franklin; and when they found employment during seventeen months un-
remittingly, in observing and recording, every hour, the variations of the
magnetical and concomitant natural phenomena, on that dreary and in-
hospitable icebound shore. The selection of the observations at Toronto,
for the purpose of comparison, is a proof as well of the accuracy of the
observations themselves, as of the value of Toronto as a place for an Observatory; and we may congratulate ourselves, that the Provincial Government resolved on recommending to the Canadian Legislature, and that the
latter most liberally responded to the recommendation, to continue the
Observatory as a Provincial establishment, placing the financial responsibility and the general oversight, under the control of the Senate of the
University of Toronto.

The Canadian people, by whom the advantage of the electric telegraph
is so thoroughly understood and appreciated, cannot fail to remember that
this,—one of the most surprising, as well as the most useful boons, which
the application of modern science has bestowed upon mankind,—was
dependent upon the discovery of those laws of electricity and magnetism
which are being further evolved, by the means of such careful and unbroken notings of varying phenomena as have been for years recorded at
our own Magnetic Observatory.

Such an establishment is worthy of the rising character of this fast-
growing community, and affords to foreign countries one of the best
proofs of our real advancement. Our progress and improvement have been
wrung from a soil which, however fertile, was covered with a dense and
pathless forest; and the toil necessary to reclaim it left to the laborer little
force, and even less of time and opportunity, for mental cultivation. It
cannot, therefore, be a matter of surprise that attempts at intellectual
progress should have tarried for the material progress which has been so
successfully achieved; that efforts to cultivate the sciences, the aesthetic
arts, the abstract philosophy, in which consist the true elements of national greatness, should but recently commence, and by degrees occupy
the thoughts and attention of the people; and it is in this view that the
Toronto Magnetic Observatory becomes a subject of honest congratula-

tion. It is a thing of a world-wide character, designed to co-operate with all other nations engaged in similar researches, and founded in the most generous spirit of philanthropy, which seeks to benefit as well future generations as our own: by the accumulation of truths, the full development and practical application whereof will only be known and made available to those who come after us, to fill our places in this busy world.

H.Y. Hind, 'Methods to be pursued in determining the data for the basis of the maps and reports of this [Assiniboine and Saskatchewan] exploration', *British North America Reports of Progress* (London, 1860), pp. 204-5.

In order to determine, within the limited period allotted for field operations, the topographical and geological character of the region indicated for exploration, and to describe faithfully and in detail its characteristic features and adaptability for settlement, it is necessary that the most expeditious method of conducting the exploratory survey be adopted, combined at the same time with every possible accuracy. As it may become advisable during the progress of the exploration to form different divisions, the following rules and suggestions are designed for general guidance, in order that the explorations and surveys may be made on a uniform system. An extensive equipment of instruments may not be supplied to each observer; he must therefore make the best use of those with which he is provided, and follow those rules which are best adapted to his mode of travelling.

Observations for latitude and longitude should be made whenever there is an opportunity, and especially at such places as the Honourable Hudson Bay Company's forts, the mouths, forks, and sources of rivers, the extremities of lakes, and at prominent hills. The magnetic variation should, if possible, be determined at every convenient camp. The delineation of the topography of the country between established positions is to be accomplished by track-survey. The courses, and cross-bearings to all conspicuous points, are to be taken by magnetic compass, and the intermediate itinerary distances to be ascertained by micrometer, or viameter, or by the measured and corrected velocity of the carts, canoes, or boats. With a view to make a complete reconnaissance of a considerable breadth of country, lateral traverses should be made at stated intervals on either side of the main lines of exploration.

When surveying rivers or lakes in a boat or canoe, the instruments essentially required for the track are a watch, a magnetic compass, a logline, and a sounding-line. At every bend of a river the direction of the reach in front is to be taken with the compass, and when the reach is very long the boat must be stopped in order that the course may be taken more accurately. The times of arriving at and departing from each bend, or the

ASTRONOMIC FIELD OBSERVATORY AT REVELSTOKE, B.C., 1886.

vertex of two courses, and the length of any halt upon a reach or course, are to be carefully noted. The velocity of the boat is to be determined by the log-line, with which frequent observations are to be made, particularly when any change in the rate is supposed to occur. In rivers it is first necessary to measure the velocity of the current, as it has to be added to or subtracted from the *apparent* rate of the boat, indicated by log-line before the true rate is ascertained. The depth, particularly of large rivers and lakes, is to be taken at close intervals, and the height of any water-mark above the present level. The width of the rivers is to be recorded (from measurement when possible) whenever it seems to vary. The height of the banks and flood-marks are also to be noted. The position and dimensions of islands, tributary streams, sand-bars, boulders, &c., are to be ascertained. It being very difficult to estimate correctly the fall or length of swift rapids, it will be necessary to make instrumental observations for this purpose, at least whenever it is possible to do so; and when they occur on large rivers, very particular descriptions of them, and their portages, if there are any, should be given. Accurate cross-sections of rivers, with the mean rate of current at each place, should be made as frequently as possible. Whenever it can be done, it would be most desirable, in addition to taking cross-sections and rate of current, to ascertain by levelling the fall of the rivers in some *measured* distance, as a quarter or half a mile. These observations and measurements will be of the greatest use in determining the descent in rivers whose general dimensions and rate of current are known, thereby enabling sections or profiles to be made of them hereafter. In ascertaining the rate of current, it should be measured with the log-line at certain intervals *across* the river, as it varies in different parts.

When surveying the coast of a lake, the boat or canoe should be steered in as straight a line as possible from one point or headland to another, and propelled at a uniform rate, so that the compass or log-line will not be required so often, and there will be more time for delineating the coast, taking soundings, and general observations. The positions of islands and intermediate points can be established more accurately by taking several intersecting bearings to them from points already determined on the course, which is the base-line, than by estimation, as the eye is oftentimes deceived in distances.

On land there are several ways of obtaining distances expeditiously, differing in accuracy according to the nature of the ground. In an *open, hilly* country, Rochon's micrometer-telescope is the best, but it may be found to retard progress. On *level* ground a viameter gives very accurate results; there are many occasions, however, when it cannot be used. Determining the track distances by the time and rate of travelling will probably be the method most used on this survey. The rate therefore at which the carts travel should be known as near as can be, and should be adhered to as much as possible. Three miles an hour is the average rate at which horses walk, but it can be tried occasionally by timing them on a *measured* distance. Due allowances must of course be made for undulations in the ground and the windings of the track. The position of distant hills or other conspicuous objects, and the width of valleys, should be determined by triangulation when the ground is suitable for measuring a base-line. The heights of hills or mountains, and the depths of valleys, should be computed trigonometrically when the level or barometer is not used. The names of all rivers, lakes, &c., should be ascertained from the Indians or half-breeds, and information procured from them relative to those parts not explored. The approximate positions and dimensions of lakes, rivers, hills, &c., according to the Indians and others, may be made use of in constructing a map of the country, but it should be strictly mentioned, and nothing should be laid down as a fact which has not been surveyed and examined.

In addition to the topographical, geological, and general character of the region to be explored (the nature of the soil, timber, vegetation, economic materials, &c., &c., specified in the general instructions, and of which *exact* descriptions should be given) it is unnecessary to state in detail what should be observed in the country, as everything should be noted. The field-books, of which different kinds are provided for the several methods of surveying, must be kept in such a clear manner that the notes recorded can be understood and plotted by other persons than the observer if necessary.

H.Y. Hind, 'Auroras', *Ibid.*, pp. 146-7.

On the night of October 2nd, when camped on Water-hen river, an aurora of unusual brilliancy and character, even in these regions, surprised us with the varied magnificence of its display of light and colour. A broad ring of strong auroral light nearly encircled the pole star. It possessed an undulatory motion, and continually shot forth, towards and beyond the zenith, vast waves of faint light. They followed one another like huge pulsations—wave after wave—expanding towards the south with undiminished strength and continuing many minutes at a time. Suddenly the waves ceased, the luminous belt or ring increased in brilliancy, lost its regular form, and here and there broke into faint streamers of a pale yellow colour. The streamers rapidly increasing soon reached the zenith, and finally meeting beyond it, shot forth from the luminous arc with swift motion and in rapid succession. Their colour varied from straw to pink. The display of streamers is quite common in this part of the continent. The waves are also not unfrequently seen; but none of the half-breeds or the Indians whom we saw a few days afterwards, had ever witnessed such a brilliant spectacle as the heavens presented during the early part of the night, when the immense pulsations, 14° to 20° in breadth, and expanding in their apparent ascent from east to west, rolled in tranquil, noiseless beauty, through the heavens overhead.

At 10 p.m., on the 27th of October, when camped on the shores of Lake Manitobah, near Oak Point, a half-breed awoke me to witness a crimson aurora of surprising magnificence. Unfortunately, a few clouds were flitting athwart the sky, which prevented the centre arc from being visible, but perhaps they increased the depth of the colour. The light was generally steady at the edges of the clouds. The appearance of streamers was recognized only in the clear portions of the sky and above the clouds, where the rose or crimson tints were much fainter. It reminded me of the reflection of a vast prairie on fire; the deep and crimson tints lasted for half an hour; then gave way to white and straw-coloured streamers, occasionally tinged with pale emerald green.

Coloured auroras are not unfrequently seen during the summer months, but they rarely possess the extraordinary beauty of those which have just been described. These beautiful "dancing spirits of the dead" impart a solemnity and charm to the still night, which must ever remain one of its most delightful characteristics in these regions.

Lake Huron, always attractive in calm summer weather, was peculiarly beautiful on the evening and night of the 25th of July 1857, during our first voyage to Red River, when lighted up by a magnificent aurora, as we neared the small Manitoulin Island. The auroral streamers converged beyond the zenith. Its base was marked by a very abrupt and well-defined sheet of light, from which waves and streamers rose from time to time. Masses of light moved continually from west to east, with an undulatory motion, occasionally folding and unfolding, with great regularity and distinctness of outline. A few minutes after 10 o'clock the base of the moving folds was tinted with delicate rose colour, passing, by imperceptible gradations, into faint emerald green above. The calm surface of the lake reflected these delicate colours, and the ever-varying motions of the auroral streamers and waves. The afternoon had been warm, with a fresh southwest breeze, and a thin haze in the same direction over-spreading the high shores of the Grand Manitoulin Island.

The beautiful spectacle presented by this aurora led to the description, hitherto unpublished as far as the narrator was aware, of a spectacle of extraordinary magnificence which had been witnessed by one of our fellow-travellers, a post-captain in the English Navy, who was making the tour of the Grand Lakes. This gentleman described his ascent to the summit of the Peak of Teneriffe, for the purpose of seeing the sun rise above the waters of the Atlantic from that imposing elevation. At the moment when the red light of the sun began to flash above the unruffled outline of the horizon, overcome with emotion at the splendour of the scene, he turned away to seek a momentary relief in the grey of the west: but unbounded astonishment and admiration seized him, on beholding, instead of a grey blank, a gigantic image of the Peak projected on the sky to the full height of 40°, and swiftly sinking into the ocean as the sun rose above its eastern outline.

Colonel Lefroy, in 1843 and 1844, enjoyed many excellent opportunities of witnessing auroras in Rupert's Land, at Fort Chipewyan, Lake Athabasca, latitude 58° 43′ north, longitude 105° 35′ 15″ west, and Fort Simpson, latitude 61° 51′ 7″ north, longitude 120° 5′ 20″ west.

Province of Canada, *Journals of the Legislative Assembly*, vol. VIII, (1849), Appendix M.M.M.

GREY TO ELGIN

Downing Street,
26 March 1847

MY LORD,—It has been for some time past in contemplation to build an Observatory at Quebec, for the express purpose of ascertaining and communicating time accurately to the Shipping.

The measure itself has been earnestly recommended by Captain Boxer, the Harbour Master of Quebec, by the Council of the Board of Trade of Quebec, by the late Commander of the Forces in Canada, Sir Richard Jackson, and by Professor Airy, the Astronomer Royal.

In a Report which was made by Professor Airy to Lord Stanley in July, 1844, it was stated that the reasons given by Captain Boxer for the establishment of an Observatory appeared to be most cogent. In every Port which has the same amount of commerce as Quebec, there should be provided means of obtaining time with that security for its general accuracy, which can be given only by the sanction of official authority. But it is especially desirable in a Port where physical circumstances make it so difficult for mariners to conduct successfully the ordinary operations of Nautical Astronomy, for obtaining time.

For reasons with which it is hardly necessary to trouble Your Lordship, it has hitherto been impracticable for the Commanding Royal Engineer to complete the task which has been assigned to him, of preparing a Plan and Estimate of the building, but as I am now sending to the Master General and Board of Ordnance some suggestions with which I have been favoured by Professor Airy, which may tend to facilitate the early completion of the Engineer's Estimate, I think it right to apprize Your Lordship that I incline at present to think, that the cost of erecting the proposed building, together with the charge of maintaining it, should be defrayed by the Legislature of Canada. And under that impression, I have instructed the Master General and Board to direct the Commanding Royal Engineer to lay the Estimate, when completed, before Your Lordship.

E.D. Ashe, Report of the Condition of Quebec Observatory, 1855, in Province of Canada, *Journals of the Legislative Assembly*, vol. xiv, (1856), Appendix 53.

Sir,—Before making a Report for the past year, I think it advisable to explain the reasons that induced the Provincial Government to establish an Observatory at Quebec, in order that it may be seen whether the ends for which it was built have been carried out.

It appears from the earnest recommendation of the Harbour Master of Quebec; of the Council of the Board of Trade of Quebec; of a late Commander of the Forces in Canada, Sir Richard Jackson; and of Professor Airy, the Astronomer Royal; that the Observatory at Quebec was built for the express purpose of ascertaining and communicating time accurately to the shipping; and of such consequence is it considered that Mariners may have an easy and certain mode of rating their Chronometer, that lately "Time Observations" have been established in many different parts of the world, in order that the fearful destruction of life and property by shipwreck may be less frequent.

This Observatory was built at a cost of £526 15s. 5d. sterling.

The Instruments consist of a 30-inch Transit and a 42-inch Telescope (which are lent by the Home Government), also two excellent Clocks, one by Dent, and the other by Molyneux; a Barometer and three Thermometers by Negretti and Zambra.

The duties consist in taking the Clock Stars as they pass the Meridian, and keeping the correct time to the nearest tenth of a second, and giving it each day (Sundays excepted) to the Shipping by dropping a Ball.

The Director's Salary, including allowances, is £239 4s. currency.

As an Assistant is absolutely necessary in case the Director is unable from sickness to attend, such an officer has been appointed, at a Salary of £122 10s., currency, and the duties of the Observatory are efficiently carried out by him, and thus, no Ship leaves the Port without an opportunity being given her of rating her Chronometer.

When we consider how small is the amount of science that the general run of Ship Masters are able to acquire, and then remember that a Steam Vessel stops neither in the darkest night nor yet in the densest fog, but proceeds onwards with rail-road speed, it will be clearly understood that it is of the greatest importance that Ships should have every facility in obtaining the right rate of their Chronometers; and the small amount requisite to keep up these establishments in the different parts of the world, dwindles into insignificance when compared with the valuable cargoes, and still more precious lives that are always exposed to the dangers which surround them; and none more fatal than an error in the reckoning.

It will be seen then, that the intention of the Government is fully carried out by the time being given to the shipping by dropping a ball at one o'clock, mean-time, at Observatory each day, which affords an easy way of determining the rate of Chronometers without sending them on shore.

Although the intention for which the Observatory was built, in regard to the shipping, is fully carried out, still, had the building been a little larger, so as to admit of the Principal living on the premises, it would have enabled Meteorological Observations to be taken (which are required to be registered every four hours) and by mounting an Equatorial, the establishment would be turned into a first class Observatory. This appears to be very desirable, when it is remembered that there is no Public Astronomical Observatory in Canada, (the Observatory at Toronto being Magnetic) whilst most other countries are contributing to the advancement of Astronomy.

The American Government, by exchanging upwards of one thousand Chronometers with the Observatory at Greenwich, and by the most approved Astronomical methods, have determined the difference of longitude between Harvard College, Boston, and that Observatory, with the utmost possible nicety; and when Dr. Tolderoy [sic], and Professor Jack, of Fredericton, by means of the Electric Telegraph, had obtained the longitude of Fredericton, with probably the same degree of exactness, I lost no time in asking permission to lead the B. A. Telegraph wire into the Observatory, in order that the longitude may be ascertained with an accuracy that could not be expected by any other mode, and after some delay in getting permission from the Royal Engineers, to erect poles and lead the wires across their works, all was ready in November last, for sending signals from one Observatory to the other; and on the night of the 15th November, 1855, the Fredericton Observatory commenced sending second beats from their Sidereal Clock, from the 20th second to the 50th second, and then waiting for 10 seconds; after which a single tap at the even minute was given; and this was continued for ten successive minutes. The first single tap was registered at Quebec, and the taps from the 20th to the 50th second in each minute, enabled us to ascertain the fraction of a second. . . .

The great advantage to be derived from fitting this Observatory up with a Telegraph Apparatus, is, that the longitude of the principal places in Canada can now be ascertained with an accuracy, and with far less expense than could possibly be expected from any other method in a Country so difficult of Triangulation.

And finally, I may remark that I endeavour to make the Observatory as useful as it is possible to do, with such limited means as are at my disposal; but should it be deemed advisable to enlarge the building, and to fit the establishment with instruments that are requisite for Meteorological Observations, and also with an Equatorial, I shall devote my whole time and energies to the advancement of science.

Applied science, in Cardwell's sense of the expression, as 'the actual investigation, by the methods of science, of the processes or products relevant to the "hardware" of particular industries'*, is a relatively recent innovation of Western society, one that did not achieve significant stature until the second half of the nineteenth century, when government and industrial research laboratories were founded. The practical application of the laws of science by the engineer or farmer is a distinct and much earlier activity, whose advantages were widely although not universally recognized throughout nineteenth-century Canada. Closely entwined with this view of useful science was the spread of popular science, which contributed to the incorporation of scientific knowledge into the educated culture of a community. Within the Province of Canada the utilitarian approach was at first adopted in a piece-meal and unco-ordinated fashion, while the cultural aspects of science, especially in Upper Canada, were little recognized until the middle of the century. Canada's situation was in this respect not far removed from that of England, but the example of the latter's prior industrial revolution, and the impetus behind the rise of Mechanics' Institutes, combined to imbue Canadian scientists with greater urgency and feelings of present inadequacy. Editorials in the *Canadian Journal* in the early 1850s exemplify these attitudes.

From at least as far back as the early eighteenth century, many physicians, military officers, and members of the clergy in Canada had displayed an interest in science. The only learned profession generally exempt from such an interest was that of law. With the spread of enthusiasm for popular science to other groups, such as tradesmen and artisans, the reservoir of potential amateur contributors to science was enlarged to the point where the formation of scientific societies could occur, and the earliest of these date in Canada from the second decade of the nineteenth century. 'Amateur' is today a highly ambivalent word. In the early nineteenth century there was no profession of 'scientist', and those most fruitfully active

---

*D.S.L. Cardwell, *The organisation of science in England*, 2nd ed. (London, 1972), p. 15.

in and indeed devoted to the field of science were necessarily amateurs, in the sense of lovers and devotees of science. The membership of the earliest societies was, accordingly, constituted of true amateurs, and of interested laymen. The presence among them of numerous professional physicians and soldiers is merely a reflection of the latter's education. The foundation of scientific societies coincided with the appearance in journals of the first articles dealing with scientific topics. Canadians have always been great readers of newspapers and journals, and these media provided the main vehicles for science to reach the literate classes.

Organizational and journalistic activity for science was restricted essentially to Lower Canada until nearly mid-century, when a group of surveyors, engineers, and amateurs of science formed the Canadian Institute in 1849, intending that it should become the leading centre of Canadian science. The essentially amateur constitution of these societies, and their lack of official status, made their financial position and hence their very continuation constantly precarious. The presidential address to the Institute in 1859 included a plea for more governmental support for the organization, in the best interests of science and of the community that would benefit directly from the useful applications of science.

The growing maturity of Canadian science was marked by recognition at international exhibitions abroad and by successful bids by Canadian societies to invite senior scientific organizations to Canada for annual meetings. The former achievement was largely due to the Geological Survey, whose displays at international exhibitions, notably at the Crystal Palace in 1851 and in Paris in 1855, provided striking attractions. The success of the Canadian exhibit at the Great Exhibition led to correspondence between the Canadian Institute and the Society of Arts, Manufactures and Commerce in Britain on how best to acquaint the world with the resources of British North America. The American Association for the Advancement of Science first met in Canada in 1857, on an occasion that all viewed as symbolic of great things, and Montreal's pride as host is apparent in a note from the *Canadian Naturalist and Geologist*.

The improvement in morale and the greater success of local scientific societies are exemplified in Dawson's presidential address of 1871 to the Natural History Society of Montreal. By 1876 Chauveau, former minister of education and Premier of Quebec, could look back on Canadian achievements with some pride. He clearly perceived the difference between English- and French-Canadian approaches to science, the former emphasizing utility, the latter cultural aspects of science; yet he too advocated increasing development and use of science and technology within the cultural context of Lower Canada.

The formation of the Royal Society of Canada, a truly national body, was indicative both of the progress of Canadian culture, including science, and of more appreciative mutual recognition by the two founding cultures.

The opening addresses of Dawson and Chauveau, respectively the first president and vice-president of the Society, admirably illustrate essential differences between those cultures, as well as fundamental areas of agreement. In the last decades of the century, discontent at the differences concerning science was agitated among French Canadians. The abbé Thomas-Etienne Hamel, educated in Paris, professor of physics and Rector of Laval University, was prominent in this debate. In a vehement presidential address to the Royal Society in 1887, 'Science and its enemies: journalism, the civil service, and politics,' he attempted to analyse French-Canadian disdain for scientific professions, laying the blame in obvious places.

The Royal Society attempted to give a national focus to scientific endeavour; new local societies meanwhile increased the participation of both professionals and amateurs in separate disciplines and in newly settled areas. One such group was the Manitoba Historical and Scientific Society; in the inaugural address of Charles N. Bell, president of the Society in 1889, we can see some of the activities of a society in the rapidly growing West. The very titles of such bodies imply the amalgam of science and humanities that constituted a liberal culture in nineteenth-century Canada.

A major thesis of this volume is that science is a characteristic of society, created by men within society, and existing in complex mutual interrelation with a host of other factors. Religion was, in the nineteenth century, among the most prominent of these factors, and it is therefore reasonable to seek and to examine the interactions between science and religion. Intellectual history exists within a social matrix.

Following the publication of Darwin's *Origin of Species* and contemporary reforming and sectarian commotion within the Churches, there were attempts to restore peace and avoid controversy. They generally failed, for such issues as the credibility of natural theology, the significance of biblical accounts of creation and the age of the earth, and the spiritual and biological uniqueness of man continued to agitate the mid-Victorian age. Darwin's works in particular were the focus of much argument, and Sir William Dawson, an ardent Presbyterian and no mean scientist, was an articulate and vocal critic who confronted all the critical issues. He shared and expounded the belief, widespread in Victorian Canada, that natural history is not opposed to religion at all, but indeed supports it by showing the wisdom and beneficence of the Creator. In his *Chain of Life in Geological Time*, he took Darwin to task, coolly and rationally arguing that species have been introduced at various times and that the geological record can substantiate such a claim. The Darwinian view is incorrect, not for moral or merely religious reasons, but because it simply cannot explain the facts. In another work, the *Origin of the World* (1893), Dawson attempted to reconcile genesis with geology, the six days of creation with the palaeontological and geological records. He clearly did not take the biblical account literally, but saw a profound parallelism between science and revelation. His works, although able and extremely popular, maintained their uncompromising stance long after the bulk of the established scientific community had accepted Darwin's views.

A continuing preoccupation, more immediate than religious controversy for the bulk of the populace, was material survival and advance. The practical applications of science, and the benefits that might accrue to agriculture, were eagerly considered. Agricultural chemistry was the main source of inspiration here. Tentatively explored in the eighteenth century, its birth as a recognized discipline came in the nineteenth century with the work of Liebig and his co-workers in Germany, and its adoption by the French and the English. Canadians were interested in agricultural science as early as the 1820s, and by around mid-century an agricultural journal was published in Lower Canada, Upper Canada had its Board of Agriculture, and several schools and organizations offered courses of lectures on agricultural chemistry. Napoléon Aubin, in his *Chimie Agricole mise à la portée de tout le monde* (1847), argued that agricultural prosperity, achieved through science, was essential for Canada's development of industrial strength. Such a view was supported in the *Agriculturist*, and associated with demands for an agricultural college. The University Question was at that time raging in Upper Canada, but the higher education at stake was classical and not designed to give professional aid to the mechanic or farmer.

The plea that scientific agricultural education and research should be sponsored by the government arose throughout British North America. All this, however, seemed remote and impractical to the farming community, and exponents of agricultural chemistry had a hard time persuading the farmer that he could benefit from their science. The correspondence and review columns of mid-century agricultural journals constantly return to the question of the practical uses of chemical and even geological science.

The debate was carried into the fortresses of University College, Toronto, where a professorial post in agriculture was established in 1850. The *Agriculturist* argued that practical agriculture, pursued at an experimental farm, should accompany academic lectures. This proposal found favour in the subsequent foundations at Guelph and Ottawa.

One area of scientific agriculture that made progress in Canada was entomology. James Fletcher's presidential address to section IV of the Royal Society in 1895 chronicles some of the achievements of Canadian entomologists. Some of this work was government sponsored, but a more important governmental agency was the Central Experimental Farm in Ottawa and its branches, founded in the 1880s; its first director, William Saunders, reviewed some of the Farm's accomplishments in an address to the Royal Society in 1907.

The First World War aroused British feeling that temporarily eclipsed much Canadian nationalism, a transition wherein the scientific community took active part, as the Royal Society's unanimous adoption of Rodolphe Lemieux's emotive resolution clearly shows. The role of applied science in modern warfare is only too familiar, and it was no accident that the first moves to organize scientific research on a national basis came in 1916, with the appointment of the Honorary Advisory Council for Scientific and Industrial Research in Canada, which later evolved into the National Research Council. A member of the Advisory Council, A.S. Mackenzie, gave a particularly cogent statement of these issues in an address to the Royal Society, and we end this section with almost the full text of his address.

A. THE ORGANIZATION AND ETHOS OF SCIENCE IN VICTORIAN CANADA

Canadian Journal 1(1852), 2-3.

It can scarcely be denied that the pursuit and cultivation of the Physical Sciences has made comparatively little progress in Canada, and by no means attained the established place which might have been looked for at this stage of our history. It is true that two Societies, directed more or less to this subject, have existed in Lower Canada for more than twenty years—the Literary and Historical Society at Quebec founded in 1824, and the Natural History Society of Montreal founded in 1827, but we have the highest authority for inferring that the latter at least has not as yet realized the expectations of its zealous founders, nor can the last Report of the authorities of the former, be deemed entirely satisfactory. Neither has practically exercised any influence in Upper Canada. But a short time ago, a celebrated naturalist had occasion to compare the skeleton of a recent specimen of the *Delphinus Leucas*, or Beluga, with some remains found under equivocal geological circumstances in the State of Vermont. In vain did he enquire of every collection with which he was acquainted, in America; the unwieldy rarity he sought was no where to be heard of. At last he remembered a museum in Copenhagen unrivalled for its riches in marine mammalians. With the cordial liberality of a brother philosopher, the distinguished naturalist who presides over that establishment, promptly met his request for a specimen, and the precious remains were shipped with much precaution, in a number of boxes and barrels, and duly wafted from Denmark to Massachusetts. Then, and not until then, did M.

Agassiz, the naturalist in question, become aware of the fact that the *Delphinus Leucas* under the name of the *White Whale* is one of the commonest frequenters of the Gulf of St. Lawrence, and that an easy journey to the banks of our noble river, would have placed him in possession of any number of specimens his researches might have required. Need we say that such a fact speaks volumes as to the neglect among us of those pursuits by which, not only are the productions of a country laid open to the use and enjoyment of its people, but the channels of scientific information kept also replenished with that knowledge of local peculiarities which is so indispensable to the progress of science.

We have referred above to the comparative non-success of the Elder Societies in Canada not in ignorance of the ability and intelligence with which ever since their formation, one zealous President or Secretary after another, has endeavoured to animate them to successful exertion, still less to undervalue those endeavours, but to enquire in perfect respect into the cause of a circumstance so frankly and honorably admitted by both, and the probability that the Canadian Institute of Upper Canada—the Society to whose recent organization we are about to refer, will be enabled to avoid a like result. First, then, it seems probable that the great vice of Society in America, that "eternal sabbathless pursuit of a man's fortune," so long ago denounced; which leaves to the mind neither leisure, taste or capacity, for the cultivation on which its happiness depends, has not failed in its effect here; not in reality devoting much of our time to anything more profitable, or half so delightful as the cultivation of literary or scientific pursuits, we have nevertheless grudged it to them, and have neglected the formation of those habits with which alone they are reconcileable. Natural History and Botany have been abandoned almost entirely to the members of an arduous and ill-remunerated profession, very few of whom can command the leisure or even incur the expenses essential to their active pursuit. The unwise habit of overtasking the strength and energy of those engaged in Instruction, or filling Professional Chairs, as if the mind can expatiate at large, while the body is bound to a tread-mill, has had something to do with it. Scientific pursuits can never make much progress while those who are professionally devoted to them, are debarred, whether by unfortunate necessity or illiberal pressure, the opportunities of self-improvement and private progress, which the ablest value the most.

It rather appears too, and we refer to this, because it is the evil which it has been principally sought to avoid, in the constitution of the Society just referred to, that the objects expressed by the titles Natural History Society, and Literary and Historical Society, are too special to be able to stand alone in this country at present. They do not include a multitude of objects in which much of the most active talent in the country is engaged, for example, those involved in the professions of the Engineer, the Artist, the Surveyor, the Architect, all of them represented by Societies of high standing in Great Britain, and therefore capable in their nature of extending the basis of similar bodies here. It must not be forgotten that until about the year 1810, one great Society satisfied almost the entire demand for this species of organization in London itself, we might almost say Great Britain, for the local societies were few in number and limited in character. The Geological Society, (1807;) the Astronomical Society, (1820;) the Asiatic Society, (1824;) the Geographical Society, (1831;) and a host more, are of very modern foundation; it would seem, therefore, that no such limitation of object has the sanction of previous experiment, and we may hope that an attempt to unite under one roof, and in one organization a full representation of the active mind of the community, may be more fortunate. It is unhappily true that the great prominence given to classical learning in England, and in all education framed on her models, has led to a surprising want of either knowledge of, or interest in, physical or mathematical science in English Society generally; which is best attested by the almost incredibly limited sale of scientific books and periodicals: it must be therefore expected that an English Colony will yield, at first, but a slender harvest of scientific results, whether of the nature of observation, experiment, or reasoning, and furnish but a small number of minds imbued with those tastes which produce them; but there is a fund of practical knowledge and thought, a wisdom of the workshop, the field, and the loom, in every community, which deserves, while it does not claim the honours of science. It is to this also that the Canadian Institute, and this journal as its present organ, addresses itself, and to this offers a medium not only as it is hoped of instruction, but of intercommunication and publicity. In referring, however, to the causes of the difficulty experienced by Literary or Scientific Societies in this country, it is impossible not to notice the habit of reading for amusement alone, which is fostered and fed by the cheap trash which loads the tables of our booksellers, and pervades society so generally. Until parents and teachers set themselves more strongly against this habit, not only for the injury it frequently does to the moral strength of the young, but still more universally, its destruction to the intellect, there will continue to be a waste of the best faculties, and a distaste for the most rational and elevating pursuits. We might add the want of Libraries and enquire why the Provincial University with its great endowments, has not long ago acquired something more deserving of that name. In the United States there are 234 Libraries containing from 5,000 volumes and upwards, including five that contain more than 50,000. In the same ratio to population, there should be nearly twenty such in the two Canadas. We doubt if there are half-a-dozen. However, in these matters cause and effect follow one another, in such recurring succession, a circle, so "vicious" is maintained, that it is useless to distinguish one from the other, and we simply refer to the facts to justify the assumption with which we started, that something more is wanted, and that something we believe, may be in part attained by the Incorporation of the Canadian Institute.

Canadian Journal, 1(1852), 14-16.

For many years the people of Canada have had just cause to regret, that information respecting the resources of the vast territory they possess, should have had such a limited circulation in the Mother Country. It is needless now to enquire into the minor causes of the extraordinary ignorance which but too generally prevails in England of the progress of the Canadas, and of the admirable opportunities they offer for the safe and remunerative investment of capital, or the exercise of well-applied industry. We are willing to rest satisfied with the explanation, which at the first blush suggests itself, that the commercial and industrial classes at home are so completely engaged with their present relations, that without their attention is pointedly drawn to a new field for enterprize, by authority upon which they can rely with confidence, they do not care to embark in projects which appear doubtful or visionary, through ignorance of the circumstances under which they are to be pursued.

It is with peculiar pleasure that we have now the opportunity of calling the attention of the Canadian public to the proposition of the Society of Arts, embraced in the subjoined correspondence....

It is almost unnecessary for us to urge upon our fellow countrymen the importance of availing themselves to the uttermost, of the opportunities presented by the Society of Arts, through whose agency the British people may be made acquainted, not only with our progress in the Industrial Arts, but more especially with the nature and extent of those vacant and neglected fields of enterprize in which this country abounds.

Correspondence relative to the establishment of Communication between the Society of Arts, Manufactures and Commerce (of London), and the Canadian Institute, with a view to advancing the knowledge of the resources and capabilities of Canada abroad, and of promoting information on the same subject within the Province.

GOVERNMENT HOUSE,
Quebec, 17th July, 1852.

Sir,—
I am directed by the Governor General to transmit to you as Correspond-

ing Secretary of the Canadian Institute, the enclosed copy of a letter from the Secretary to the Society of Arts, Manufactures and Commerce, to Her Majesty's Principal Secretary of State for the Colonies, with enclosures having reference to the establishment of a Correspondence between the Society of Arts, and similar institutions in the Colonies. His Excellency is desirous to ascertain, through you, whether the Canadian Institute will be disposed to engage in the proposed Correspondence with the Society of Arts, as he believes that the objects of the Institute and the interests of the Province would be promoted thereby.

I have the honor to be, Sir,

Your most obedient humble Servant,

R. BRUCE,

Gov. Secretary.

F. Cumberland, Esq.

    Corresponding Secretary,

      Canadian Institute.

Copy of a Letter from the Secretary to the Society of Arts, Manufactures and Commerce, to Her Majesty's Principal Secretary of State for the Colonies.

Society of Arts, John Street Adelphi, London,

26th March, 1852.

Sir,—

I am directed by the Council of the Society of Arts to acquaint you, that they have appointed a Committee . . . to consider the best means of making the Society useful in advancing the knowledge of the resources and capabilities of the numerous British Colonies in all quarters of the world, and in furnishing the Colonies themselves with such information as may be required on subjects connected with Arts, Manufactures and Commerce.

The accompanying Enclosures, Nos. 1 and 2, will explain the Constitution of the Society, the objects they have in view in adopting the present measure, and the means which they possess of carrying them into effect.

The Council conceive that one of the first steps towards the attainment of their Objects, will be the establishment of a Correspondence with similar Institutions in the Colonies; or, in the smaller Colonies, where no such Institutions exist, with a Committee consisting of three or more Members, in all cases, where volunteers for such a purpose can be found.

I am therefore, to express the hope of the Council, that you will be pleased to accord to the Society the advantages of that co-operation and assistance which the Colonial Office is so well able to afford, to enable them to place themselves thus in correspondence with the numerous

Colonies. And, as the readiest means of doing so, I am directed to transmit to you Printed Copies of the present Letter and the Enclosures, which the Council trusts you will have the goodness to forward to the Governors of Colonies, with such instructions for their judicious distribution as may appear best calculated to ensure their practical utility.

I have the honor to be, Sir,

Your most obedient Servant,

GEORGE GROVE,

Secretary.

ENCLOSURE NO. 1

Brief Statement of the Objects . . . of the Society for the Encouragement of Arts, Manufactures, and Commerce:

Objects:—The Society for the encouragement of Arts, Manufactures and Commerce was founded in 1754, and incorporated under the above name by Royal Charter in 1847, they are summed up in the Charter as—"Generally to assist in the advancement, development and practical application of Science in connection with the Arts, Manufactures and Commerce of the Country." . . .

ENCLOSURE NO. 2

The principal objects which the Council have in view in establishing the Colonial Committee may be generally enumerated under the following heads:—

1. To make known to the Mercantile and general Public of this Country the principal products of each of the Colonies, and the facilities for obtaining them.

2. To point out to the Colonists any of those Products which may be advantageously imported into England.

3. To afford such information as any Colony may require in regard to Implements, Machinery, Chemical or other processes necessary to the prosecution of its special branches of Industry.

4. To exhibit and make known to the British Public, Inventions which Colonists have otherwise great difficulty in introducing into notice, that being one of the principal branches of the Society's operations.

5. To collect for the Thirty Standing Committees, information relative to the various departments of Trade in the Colonies.

6. To make a comparison of Coins, Weights and Measures, as used in the Colonies, and to receive and discuss propositions for giving them uniformity.

7. To investigate and report upon the operations of the Patent Laws in the Colonies.

It is hoped that the periodical transmission of the printed Proceedings of the Society of Arts may often convey valuable information to distant Colonies, and the Society hope to enrich their own Annual Volume by communications from kindred Associations, and from Individuals in the Colonies.

The Council feel confident that these measures cannot fail to be of use both to the Mother Country and to the Colonies, and that should they be unsuccessful in some of the objects above enumerated, benefit will ensue from the remainder.

It may be desirable here to state the reasons which induce the Council to originate the present scheme.

It was as President to the Society of Arts, that His Royal Highness Prince Albert first announced to the World the project of the Exhibition of 1851. The Society had a considerable share in the early progress of the Exhibition, and counts amongst its Members a large proportion of those who took an active part in that great Work.

The Society also contains many Members eminent in the several branches of science, and influential in the Country, and consequently the Society possesses the means of making extensively known, amongst the Manufacturers and Public of Great Britain, any new or important products which may be made available in the Arts, Commerce, or Manufactures of the Country. As a recent instance of this nature, it may be mentioned that Gutta Percha and its valuable properties were made known through the exertions of the Society.

The Correspondence that has taken place with the Colonies, on account of the Exhibition, has brought to notice that those by whom it has been conducted are capable of affording a vast amount of information, which only requires to be collected and printed, to make it of great use to this Country. And the anxiety which has been evinced for such information as, it is hoped, may be advantageously furnished by Members of the Society, has directed attention to the fact that they have now no direct means of obtaining such information. The Society feels confident, that those who took an active part in the promotion of the Exhibition, will be the first to come forward and render assistance to any scheme such as the present, by which efforts are made to perpetuate its results.

It may be interesting also here to refer to a few of the advantages which have been actually derived from the display of Colonial Produce at the Great Exhibition.

Isinglass had hitherto been regarded as obtainable principally from the fish of the Russian rivers. But it has been ascertained that the rivers of Canada abound with fish producing Isinglass of the first quality, and that a new industrial occupation is thus open to the Canadians, whilst a supply of Isinglass can be furnished to this country at a much more reasonable price than hitherto.

Another remarkable instance is the discovery that Corundum, which has served many of the purposes of diamond and emery powder in India for a long period, might also be brought into use in this country; a mineral with which it is believed but a very small portion of the British public had hitherto been acquainted, and which it is suspected has in some instances been sold to our large firms under the name of Diamond powder. . . .

Notwithstanding that these and other substances have been brought into notice, Colonial Produce was on the whole but indifferently represented in the Exhibition, and the Council confidently hope that the means they have now adopted may lead to the formation, at some future period, of a permanent Exhibition of Colonial Produce, either separately, or what would perhaps be preferable, as part of The Collection arising out of the Great Exhibition, from the exertions of The Royal Commissioners.

(Signed)

GEORGE GROVE,
Secretary Society of Arts.

CANADIAN INSTITUTE,
Toronto, 31st July, 1852.

Sir,—

I, have the honor to acknowledge the receipt of your Letter of the 17th instant, with enclosures transmitted by command of His Excellency the Governor General, having reference to the establishment of a Correspondence between the Society of Arts and the Canadian Institute, for certain purposes connected with Arts, Manufactures and Commerce, therein set forth, and in reply to inform you, that having submitted the same to the Council of the Canadian Institute, I am directed to request that you will assure His Excellency that the Council will gladly take every means in its power of promoting the intentions of the Society of Arts; that it will be happy to receive any communications and act upon the suggestions of that Society; and is prepared to become the medium of transmission to it of information relative to the production and resources of Canada; of the inventions of persons resident in the Province, together with whatever else of local interest may appear to fall within the scope of its enquiries, or be deserving of its notice.

I have the honor to be, Sir,
Your most obedient humble Servant,

FRED. CUMBERLAND,
Corresponding Secretary.

THE HON. R. BRUCE,
Governor's Secretary,
Quebec.

Canadian Naturalist and Geologist, 2(1857), 241-3.

ELEVENTH MEETING OF THE AMERICAN ASSOCIATION
FOR THE ADVANCEMENT OF SCIENCE

The American Scientific Congress has just closed its sitting in our good Canadian City of Montreal, effecting a virtual Scientific annexation of American to British American minds, in their action in the wide field of physical and natural science. The Meeting has been a highly successful one, creditable alike to the assembled science of America, and to the city of Montreal, and we trust will produce lasting good effects, unalloyed by those small jealousies and heartburnings that too often remain after great assemblies of men, whatever the objects in which they may have been engaged.

We cannot attempt to present to our readers a full report of the proceedings of the Association. In its more ephemeral form this has been well done by the newspapers, and in its more permanent form it will appear in due time in the proceedings. We purpose merely to preserve a record of leading features of scientific progress evidenced in the meeting, and of points especially important to Canada. In doing so we shall draw largely on the reports in the Montreal newspapers, especially the *Herald* and *Gazette*; reports which are in the highest degree creditable to those prints and could not easily have been surpassed in any part of America.

The opening Meeting was imposing and interesting. The Divine blessing was invoked on the proceedings by the Bishop of Montreal, in a singularly appropriate and beautiful prayer. Prof. Caswell, the Vice-President, who, owing to the lamented death of the great microscopist Bailey, was called to preside, delivered a short but happily conceived inaugural address, from which we extract the following sentences, as worthy of being embalmed in the memory of Canadians and Americans:—"It augurs well for the interests of Science that so many have come to this gathering to place their choicest contributions on her altar, and to welcome to her fellowship the humblest laborer in her cause. I think also that it is a matter of congratulation that we have met without the limits of the United States. However it may have been in former times, it is not now the case that mountains or seas interposed make enemies of nations. In the onward march of Science, it is one of the felicities of our time, that little account is taken of the boundaries that separate states and kingdoms. The discoverer of a new law or principle in nature, of a new process in the arts or a new instrument of research, of beneficial tendency, is speedily heralded over land and ocean; is welcomed as the benefactor of his race, and is immediately put into communication with the whole civilized world. We have before us a practical illustration of the amenities of science. We of the United States are here convened on British soil, little thinking that we have passed the boundary of the protection of American law, or that amidst the generous hospitality of this enterprising commercial capital of a noble Province of Great Britain, we are aliens to the British constitution. We have left the American eagle, but we assure the gentlemen of Canada that we feel in no danger of being harmed by the British lion. I have said that we are aliens to the British constitution; but that must of course be taken in the narrowest and most technical sense, for I am proud to say, on deliberate conviction, that nothing is alien to the British Constitution that looks to the perfection of knowledge, to the furtherance of the arts or the amelioration of the condition of humanity. I further say, and (turning to Gen. Eyre) I here speak by permission, that the proudest achievements of British arms, and they have been proud enough for the highest desires of ambition or of glory, have been less glorious than that patronage of science, that success in the arts, and those attempts to improve the condition of our race, which have placed Old England in the van of nations. At no period of time has that patronage been more wisely directed, or those noble efforts more earnestly persevered in than under the reign of the present illustrious Queen, whose virtues are alike the ornament of her sex and Crown. There is something of special fitness in our assembling here at this time—at a moment when England and America are shaking hands with each other across the broad bosom of the Atlantic, when that electric chain which is to bind them in perpetual friendship, is being placed securely in the depths of the ocean far out of reach of any temporary storms which may impair its repose or lessen its efficiency."

The Association was then welcomed to Canada, on behalf of the Province, the Local Committee, and the National History Society, in few but fitting words, by His Excellency the Administrator of the Government, Sir W. E. Logan, and Principal Dawson.

The division into sections is not so perfect in the American as in the British Association. The smaller number of scientific men and of papers, affords a reason for this; but we think that much more and better work could be done by a more minute sub-division. At the late meeting, after the primary division, established by the constitution, into sections of Physical and Natural science, but one sub-section was formed, that of Ethnology and Statistics. We shall take the matters presented to these sections in their order; dwelling especially, however, on the subjects more appropriate to the sphere of this publication.

Canadian Journal, NS 4 (1859), 95-6.

PRESIDENTIAL ADDRESS

Passing from this subject, I am naturally led to the consideration of our position in respect to the aid which we receive in the shape of pecuniary support from the Provincial Government; and while I would record the grateful sense which I, in common, I am sure, with all my fellow-members, entertain of the long continued liberality of the Government towards us, I cannot help expressing my regret that it has been deemed advisable to withdraw the grant formerly made to the Toronto Athenæum, and which we have enjoyed since the amalgamation of the two institutions. The depressed state of the financial affairs of the country, and the consequent necessity for a strict economy in the expenditure of the public moneys, induced the Government, I suppose, to limit their liberality last year to the grant of £250, but I hope that it will not be found necessary to confine it to that amount hereafter.

I cannot but regard a liberal appropriation from the public funds for the purpose of aiding and supporting societies having for their objects the advancement of science and the spread of knowledge, as a wise and judicious act on the part of any government, and with reference to this point I trust you will pardon my . . . quoting from the excellent address of Professor Owen. . . . In that part of it in which he alludes to the aid and countenance which the British Government had always given to science and scientific institutions, he proceeds to show how science makes *return* to governments for fostering and aiding her endeavours for the public weal:

"Every practical application of the discoveries of science," says the Professor, "tends to the same end as that which the enlightened statesman has in view. The steam engine in its manifold application, the crime-decreasing gas lamp, the lightning conductors, the electric telegraph, the law of storms, and rules for the mariner's guidance in them, the power of rendering surgical operation painless, the measures for preserving public health, and for preventing or mitigating epidemics—such are among the more important practical results of pure scientific research, with which mankind have been blessed, and states enriched. They are evidence unmistakeable of the close affinity between the aims and tendencies of science, and those of true state policy. In proportion to the activity, productivity, and prosperity of a community, is its power of *responding to the calls of the Finance Minister*. By a far seeing one, the man of science will be regarded with a favorable eye, not less for unlooked for streams of wealth that have already flowed, but for those that may in future arise out of the application of the abstract truths, to the discovery of which he devotes himself."

The Canadian Naturalist and Quarterly Journal of Science, 6(1871), 2-5.

ANNUAL ADDRESS OF THE PRESIDENT OF THE
NATURAL HISTORY SOCIETY OF MONTREAL,
PRINCIPAL DAWSON, LL.D., F.R.S.,
Delivered May 19th, 1871.

The scientific papers presented to the Society in the past year have been numerous and valuable and most of them have been printed in full in our journal, the *Canadian Naturalist*. The following may be especially mentioned: "Aquaria Studies," Part 2d, by Mr. A. S. Ritchie; "On a specimen of *Beluga* recently discovered at Cornwall, Ontario," by E. Billings, Esq., F.G.S.; "On the Earthquake of October 20th, 1870," by Principal Dawson, F.R.S.; "On Canadian Phosphates, in their application to Agriculture," by Gordon Broome, F.G.S.; "On the Origin of Granite," by G. A. Kinahan, Esq., of Dublin; "Notes on Vegetable Productions,; by Major G. E. Bulger; "On the species of Deer inhabiting Canada," by Prof. R. Bell, F.G.S.; "On the Sanitary Condition of Montreal," by Dr. P. P. Carpenter; "On the Foraminifera of the Gulf and River St. Lawrence," by G. M. Dawson; "On Canadian Foraminifera," by J. F. Whiteaves, F.G.S.; "On some New Facts in Fossil Botany," by Principal Dawson, F.R.S.; "On the occurrence of Diamonds in New South Wales," by Mr. Norman Taylor, and Prof. A. Thompson; communicated by A. R. C. Selwyn Esq., F.G.S.; "On the Structure and affinities of the Brachiopoda," by Prof. Morse; "On a Mineral Silicate injecting Palæozoic Crinoids," by Dr. T. Sterry Hunt, F.R.S.; "On the Origin and Classification of Crystalline Rocks," by Mr. Thomas Macfarlane; "On the Plants of the West Coast of Newfoundland," by John Bell, M.A., M.D.; "On Canadian Diatomaceæ," by Mr. W. Osler; "On the Botany of the Counties of Hastings and Addington," by B. J. Harrington, B.A.

Beside these, we have reprinted in the *Naturalist* several important papers by Dr. Hunt, Mr. Billings, and others, with the view of making them more fully known to students of nature in Canada.

ERRONEOUS PUBLIC OPINIONS

Of the scientific value of these papers, and of the amount of original work which they evince, it is unnecessary that I should speak; but it is sometimes alleged that societies of this kind are of no practical utility; that their labours are merely the industrious idleness of unpractical dreamers and enthusiasts. Nothing could be more unjust than such an assertion. Science, cultivated for its own sake, and without any reference to practical application, is a noble and elevating pursuit, full of beneficial influence on mental culture, and by the training which it affords, fitting men for the practical business of life better than most other studies. Further, it is by this disinterested pursuit of science, for its own sake, that many of the most practically useful arts and improvements of arts had their birth. Besides this, most of the investigations of the naturalist have a direct bearing on utilitarian pursuits. In illustration of this statement I need go no further than our own last volume. An eminent example is afforded by the paper of Mr. Gordon Broome on Canadian phosphates. Here we have set before us three pregnant classes of facts: First—Phosphates are essential ingredients of all our cultivated plants, and especially of those which are most valuable as food. In order that they may grow, these plants must obtain phosphates from the soil, and if the quantity be deficient so will the crop. Of the ashes of wheat, 50 per cent consist of phosphoric acid, and without this the wheat cannot be produced; nor if produced would it be so valuable as food. Second—The culture of cereals is constantly abstracting this valuable substance from our soils. The analyses of Dr. Hunt have shown long ago that the principal cause of the exhaustion of the worn-out wheat lands of Canada is the withdrawal of the phosphates, and that fertility cannot be restored without replacing these. In 292,533 tons of wheat and wheaten flour exported from Montreal in 1869, there were, according to Mr. Broome, 2,340 tons of phosphoric acid, and this was equal to the total impoverishment of more than 70,000 acres of fertile land. To replace it would require, according to Mr. Broome, 5,850 tons of the richest natural phosphate of lime or 13,728 tons of super-phosphates as ordinarily sold, at a cost of more than $480,000. These facts become startling and alarming when we consider that very little phosphoric acid in any form is being applied to replace this enormous waste. Yet so great is now the demand for these manures that super-phosphates to the value of $8,750,000 are annually manufactured in England from mineral phosphate of lime, beside the extensive importations of bones and guano. Third—Canada is especially rich in natural mineral phosphates, as yet little utilized, and might supply her own wants, and those of half the world beside, if industry and skill were directed to this object.

Putting these three classes of facts together, as they are presented by Mr. Broome, we have before us, on the one hand, an immense abyss of waste, poverty and depopulation yawning before our agricultural interests; and on the other, inexhaustible sources of wealth and prosperity lying within reach of scientific skill, and the conditions necessary to utilize which were well pointed out in the paper referred to. It is true that these facts and conclusions have been previously stated and enforced, but they remain as an illustration of scientific truths of important practical value still very little acted on.

Naturalists are sometimes accused of being so foolish as to chase butterflies, and the culture of cabbage is not usually regarded as a very scientific operation; yet any one who reads a paper on the Cabbage butterfly read at one of our meetings by the late Mr. Ritchie, may easily discover that there may be practical utility in studying butterflies, and that science may be applied to the culture of the most commonplace of vegetables. A valuable crop, worth many thousands of dollars, is hopelessly destroyed by enemies not previously known, and appearing as if by magic. Entomology informs us that the destroyer is a well known European insect. It tells how it reached this country and that it might have been exterminated by a child in an hour on its first appearance. But allow it to multiply unchecked, it soon fills all our gardens and fields with its devastating multitudes, and the cultivators of cabbages and cauliflowers are in despair. But Entomology proceeds to show that the case is not yet hopeless, and that means may still be found to arrest its ravages.

Unfortunately, we have as yet no public official bureau of Entomology, and therefore we must be indebted for such information to men who, like our late associate Ritchie, snatch from arduous business pursuits the hours that enable them thus to benefit their country. Ontario is in advance of us in this, and has in the present year produced an important contribution to practical science in the report of the Fruit Grower's Association, which includes, among other matters, three papers on applied Entomology; that on Insects affecting the Apple, by Rev. C. [J.] S. Bethune; that on Insects affecting the Grape, by Mr. N. Saunders; and that on Insects affecting the Plum, by Mr. E. B. Reed. These are most creditable productions and of much practical value.

I would mention here that though we have among us several diligent and successful students of insects, yet we have no one at present who has taken up the mantle of Mr. Ritchie as a describer of their habits. I trust that some of our younger members will at once enter on this promising and useful field.

WORK DONE

Looking at the amount of work done by our Society in the course of the year, I think it will bear comparison with that of similar societies elsewhere. We have not before us so large an amount of matter as that accumulated by the great central societies of the Mother Country and the United States; but we exceed in this respect most of the local societies of Great Britain, outside of London, and most of those in America with the exception of a few of the more important. With regard to the quality of scientific matter, we can boast many papers of which any society might gladly take the credit, while all of the papers which we publish are at least of local value and importance. This Society is, on this account, now recognized as the chief exponent of Canadian Natural History, and its journal is sought by all interested in the aspects of nature in this part of America. The responsibility which devolves upon us in this aspect of our work, is,

I think, worthy of our consideration, with reference to our future operations, and to this subject I would desire to devote the remainder of this address.

One of our functions as a local society I think we have well and efficiently performed. It is that of accumulating and arranging for study the natural productions of this country. Our collections of mammals, birds, insects and mollusks of Canada are now nearly complete up to the present state of knowledge, and we have also valuable collections in other departments of Zoology. Our curator, Mr. Whiteaves, has done very much to give to these collections a scientific value by careful and accurate arrangement.

P.-J.-O. Chauveau, *Instruction Publique au Canada* (Quebec, 1876), pp. 311-14. (Translation.)

LITERARY AND INTELLECTUAL MOVEMENT

We promised to speak, in closing, of the literary and intellectual movement among us; . . . The considerable number of classical institutions, early established in the two large provinces, has favoured the progress of literature and science; but without a doubt, it seems that, by one of those natural affinities of which we spoke, the English population has been more inclined toward mathematical, physical, and natural sciences, and the French population towards moral and political science, history, literature, and the fine arts.

In a new country where there are few great fortunes, where the population is not very dense, where the professions that we call *liberal* are the refuge and patrimony of nearly every educated man, we cannot expect to find that science, literature, and fine arts are cultivated for anything but amusement, a pastime, a means to renown and celebrity. Until just a few years ago, the publication of a book or even a pamphlet was more often a source of expense to the author than a source of profit.

The English population, more fortunate than the French, has continually received from Europe a contingent of educated men, writers, and journalists; the literati and learned men born in Europe are as numerous as those who saw the light of day in Canada. Among these latter are found two geologists who have obtained a European reputation, Sir William Logan, born in the province of Quebec, and Professor Dawson, born in Nova Scotia, as was that well-known literary figure, Judge Haliburton, the author of *Sam Slick*.

Since the conquest, the French population has been left, in this regard as in all others, to its own vigour and its own resources; this is a fact that many of its detractors have not always taken into account; the longer we were separated from France, the more it became necessary for educated men to learn and to speak two languages, which was a great obstacle for us. The influence of all these circumstances upon our youthful literature has been well set out by Hector Fabre, in a work published in the *Transactions* of the Literary and Historical Society of Quebec.

If one may judge an intellectual movement simply by the number of newspapers and public libraries, the Province of Ontario is the decided leader over all the other provinces. We have been able to see the considerable number of volumes that have been placed in its municipal and school libraries; as for newspapers, they are legion. However, the great daily papers in Toronto seem to dominate over the teeming local press and they even have a considerable circulation in the other provinces.* The periodical and purely literary or scientific press is far from being so flourishing; the attempts that have been made in this genre, from time to time, have not yet enjoyed the same success as those that have been reviewed with great perseverance, since the beginning of the century, in Lower Canada. Anglo-Canadian literature not only has to submit to competition from English literature from Europe; it also has competition from English literature in the United States, from a cheaper press, from illustrated magazines — so numerous in our neighbours' land — and from their reprinting of English works. Our railways, our hotels, our steam packets are inundated by their products. Although the French-Canadian book market has naturally been invaded by French publications from Europe, French-Canadian productions are daily taking a larger and more important place in Europe. . . .

Proceedings and Transactions of the Royal Society of Canada, 1 (1882), pp. vi-xi.

J.W. DAWSON, PRESIDENTIAL ADDRESS

My Lord and Gentlemen; Ladies and Gentlemen,—We meet to-day to inaugurate a new era in the progress of Canadian literature and science, by the foundation of a body akin to those great national societies which in Great Britain and elsewhere have borne so important a part in the advancement of science and letters. The idea of such a society for this country may not be altogether new; but if broached at all, it has been abandoned from the inability of its advocates to gather together from our widely distributed provinces the elements necessary to its success. Now it presents itself under different and happier conditions. In the mother country, the reign of Queen Victoria, our gracious Sovereign, has been specially marked by the patronage of every effort for the growth of education, literature, science and art, not only on her part but on that of the

---

*Several German newspapers and a French newspaper are published in the province of Ontario.

lamented Prince Albert and of the members of the Royal family. It is fitting that here too the representative of Royalty should exert the same influence and our present Governor-General has undoubtedly a personal as well as a hereditary right to be the patron of progress and culture in literature and science. Besides this, political consolidation and improved means of intercourse have been welding together our formerly scattered provinces and causing much more intimate relations than formerly to subsist between men of letters and of science.

We are sometimes told that the enterprise in which we are engaged is premature, that, like some tender plant too early exposed to the frost of our Canadian spring, it will be nipped and perish. But we must remember that in a country situated as this is nearly everything is in some sense premature. It is with us a time of breaking-up ground and sowing and planting, not a time of reaping or gathering fruit, and unless this generation of Canadians is content, like those that have preceded it, to sow what others must reap in its full maturity, there will be little hope for our country. In Canada at present, whether in science, in literature, in art or in education, we look around in vain for anything that is fully ripe. We see only the rudiments and the beginnings of things, but if these are healthy and growing, we should regard them with hope, should cherish and nurture them as the germs of greater things in the future. Yet there is a charm in this very immaturity, and it brings with it great opportunities. We have the freedom and freshness of a youthful nationality. We can trace out new paths which must be followed by our successors; we have the right to plant wherever we please the trees under whose shade they will sit. The independence which we thus enjoy, and the originality which we can claim, are in themselves privileges, but privileges that carry with them great responsibilities.

Allow me to present to you a few thoughts bearing on this aspect of our position, and, in doing so, to confine myself chiefly to the side of science, since my friend, Dr. Chauveau, who is to follow, is so much better able to lay it before you from the literary point of view.

Young though our country is, we are already the heirs of the labours of many eminent workers in science, who have passed away or have been removed from this country. In geology, the names of Bigsby, Bayfield, Baddeley, Logan, Lyell, Billings, Hector and Isbister, will occur to all who have studied the geological structure of Canada, and there are younger men like McOuat and Hartley, too early snatched away, who have left behind them valuable records of their labours. In botany and zoology we can point to Michaux, Pursh, Hooker, Shepherd, Bourgeau, Douglas, Menzies, Richardson, Lord and Brunet. These are but a few of the more eminent labourers in the natural history of this country, without mentioning the many living workers who still remain to it; and were it the object of this Society merely to collect and reproduce and bring up to date

SIR SANDFORD FLEMING

what these older men have done, it would have no small task before it. But to this we have to add the voluminous reports of the Geological Survey, and the numerous papers and other publications of the men who are still with us. In natural science we thus have a large mass of accumulated capital on which to base our future operations, along with an unlimited scope for further researches.

The older men among us know how much has been done within the lifetime of the present generation. When as a young man I began to look around for means of scientific education, there was no regular course of natural science in any of our colleges, though chemistry and physics were already taught in some of them. There were no collections in geology or natural history except the private cabinets of a few zealous workers. The Geological Survey of Canada had not then been thought of. There were no special schools of practical science, no scientific libraries, no scientific publications, and scarcely any printed information accessible. In these circumstances, when I proposed to devote myself to geological pursuits, I had to go abroad for means of training not then equal to that which can now be obtained in many of our Canadian colleges. Nor at that time were there public employments in this country to which a young geologist or naturalist could aspire. It is true this was more than forty years ago, but in looking back it would seem but as yesterday, were not these years marked by the work that has been done, the mass of material accumulated and the scientific institutions established within that time. Those who began their scientific work in these circumstances may be excused for taking somewhat hopeful views as to the future.

Perhaps at present the danger is that we may be content to remain in the position we have reached, without attempting anything farther; and, however inconsistent this may be, it is easy to combine the fear that any movement in advance may be rash and premature, with the self-satisfied belief that we have already advanced so far that little remains to be attained. We must bear in mind, however, that we have still much to do to place us on a level with most other countries. With the exception of the somewhat meagre grants to the Geological Survey and to the Meteorological Service, the Government of Canada gives nothing in aid of scientific research. What is done for scientific education by local societies must, under our system, be done by the separate Provinces, and is necessarily unequal and imperfect. Few large endowments have been given for scientific purposes. We have had no national society or association comparable with those in other countries. Yet we are looking forward to a great future. Wealth and population are moving rapidly onward, and the question is whether culture of the higher grade shall keep pace with the headlong rush of material progress. Various elements may enter into the answer of this question, but undoubtedly the formation of such a society as this is one of these of the utmost importance; and, even though at the present time the project may fail of success or be only partially effective (of which, however, I have no apprehension), it must be renewed till finally enabled to establish itself.

Another consideration bearing on this question is the vastness of the territory which we possess, and for the scientific development of which we have assumed the responsibility. Canada comprises one-half of the great North American continent, reaching for three thousand miles from east to west, and extending from south to north from the latitude of $42°$ to the Polar Sea. In this area we have representatives of all the geological formations, from the Laurentian and Huronian, to which Canada has the honour of giving names, the Post-pliocene and modern. Of some of these formations we have more magnificent developments than any other country. In zoology our land area extends from the home of the musk-ox on the north to that of the rattlesnake on the south, and we have perhaps the greatest area possessed by any country for the study of fresh water animals. Our marine zoology includes that of the North Atlantic, the North Pacific and the Arctic Ocean. In botany we have the floras of the Atlantic and Pacific slopes, of the western plains and of the Arctic zone. In physical, astronomical and meteorological investigations we have the advantage of vast area, of varied climate and conditions; while these circumstances in themselves imply responsibilities in connection with the progress of science not here only but throughout the world. Much is no doubt being done to cultivate these vast fields of research, and I would not for a moment underrate the efforts being made and the arduous labours, perils and privations to which the pioneers in these fields are even now subjected, but what is being done is relatively insignificant. Many letters from abroad reach me every year asking for information or reference to Canadian workers in specialties which no one here is studying; and I know that most of our active naturalists are continually driven by such demands to take up new lines of investigation in addition to those already more than sufficient to occupy their time and energy. Were it not for the aid indirectly given to us by the magnificent and costly surveys and commissions of the United States, which freely invade Canadian territory whenever they find any profitable ground that we are not occupying, we should be still more helpless in these respects. Is there not in these circumstances reason for combination of effort, and for the best possible arrangements for the distribution of our small force over the vast area which it has to maintain.

I have dealt sufficiently long on topics which indicate that the time has fully come for the institution of the Royal Society of Canada. Let us turn for a moment to the consideration of the ends which it may seek to attain and the means of their attainment.

I would place here first the establishment of a bond of union between the scattered workers now widely separated in different parts of the

Dominion. Our men of science are so few and our country so extensive that it is difficult to find in any one place or within reasonable distance of each other, half a dozen active workers in science. There is thus great lack of sympathy and stimulus, and of the discussion and interchange of ideas which tend so much to correct as well as to encourage. The lonely worker finds his energies flag, and is drawn away by the pressure of more popular pursuits, while his notions become one-sided and inaccurate through want of friendly conflict with men of like powers and pursuits. Even if this Society can meet but once a year, something may be done to remedy the evils of isolation.

Again, means are lacking for the adequate publication of results. True we have the reports of the Geological Survey, and Transactions are published by some of the local societies, but the resources at the disposal of these bodies are altogether inadequate, and for anything extensive or costly we have to seek means of publication abroad; but this can be secured only under special circumstances; and while the public results of Canadian science become so widely scattered as to be accessible with difficulty, much that would be of scientific value fails of adequate publication, more especially in the matter of illustrations. Thus the Canadian naturalist is often obliged to be content with the publication of his work in an inferior style and poorly illustrated, so that it has an aspect of inferiority to work really no better, which in the United States or the mother country has the benefit of sumptuous publication and illustration. On this account he has often the added mortification of finding his work overlooked or neglected, and not infrequently while he is looking in vain for means of publication, that which he has attained by long and diligent labour is taken away from him by its previous issue abroad. In this way also it very often happens that collectors who have amassed important material of great scientific value are induced to place it in the hands of specialists in other countries, who have at their command means of publication not possessed by equally competent men here. The injury which Canadian science and the reputation of Canada sustain in this way is well known to many who are present and who have been personal sufferers.

Should this Society have sufficient means placed at its disposal to publish Transactions equal — I shall not say to those of the Royal Society of London or the Smithsonian Institution at Washington — but to those of such bodies as the Philadelphia Academy or the Boston Society of Natural History, an incalculable stimulus would be given to science in Canada, by promoting research, by securing to this country the credit of the work done in it, by collecting the information now widely scattered, and by enabling scientific men abroad to learn what is being done here. It is not intended that such means of publication should be limited to the works of members of the Society. In this respect it will constitute a judicial body to decide as to what may deserve publication. Its Transactions should be open to good papers from any source, and should in this way enable the younger and less known men of science to add to their own reputation and that of the country, and to prepare the way for admission to membership of this Society.

Few expenditures of public money are more profitable to the State than that which promotes scientific publication. The actual researches made imply much individual labour and expense, no part of which falls on the public funds; and by the comparatively small cost of publication the country obtains the benefit of the results obtained, its mental and industrial progress is stimulated, and it acquires reputation abroad. This is now so well understood that in most countries public aid is given to research as well as to publication. Here we may be content, in the first instance, with the latter alone; but, if the Society shall at first be sustained by the Government, it may be hoped that, as in older countries, private benefactions and bequests will flow into it, so that eventually it may be able not merely to afford means of publication but to extend substantial aid to young and struggling men of science who are following out, under difficulties, important investigations.

In return for aid given to this Society, the Government may also have the benefit of its advice as a body of experts in any case of need. The most insignificant natural agencies sometimes attain to national importance. A locust, a midge, or a parasite fungus, may suddenly reduce to naught the calculations of a finance-minister. The great natural resources of the land and of the sea are alike under the control of laws known to science. We are occasionally called on to take our part in the observation of astronomical and atmospheric phenomena of world-wide interest. In such cases it is the practice of all civilized governments to have recourse to scientific advice, and in a Society like this our Government can command a body of men free from the distracting influences of private and local interests and able to warn against the schemes of charlatans and pretenders.

Another object which we should have in view is that of concentrating the benefits of the several local societies scattered throughout the Dominion. Some of these are of long standing and have done much original work. The Literary and Historical Society of Quebec is, I believe, the oldest of these bodies, and its Transactions include not merely literature and history but much that is of great value in natural science, while it has been more successful than any of our other societies in the accumulation of a library. The Natural History Society of Montreal, of which I have had the honour to be a member for twenty-seven years, is now in its fifty-third year. It has published seventeen volumes of Proceedings, including probably a larger mass of original information respecting the natural history of Canada than is to be found in any other publication. It has accumulated a valuable museum, and has done much to popularize science. It has twice induced the American Association for the Advancement of Science to hold its meetings in Canada, and was the first body to propose the establishment of a Geological Survey. The Canadian Institute of Toronto, occupying the field of literature as well as science, though a younger has been a more vigorous society; and its Transactions are equally voluminous and valuable. The Natural History Society of St. John, New Brunswick, though it has not published so much, has carried out some very important researches in local geology, which are known and valued throughout the world. The Nova Scotian Institute of Natural Science is a flourishing body and publishes valuable Transactions. The Institut Canadien of Quebec, and the Ottawa Natural History Society, are also flourishing and useful institutions. The new Natural History Society of Manitoba has entered on a vigorous and hopeful career. There are also in the Dominion some societies of great value cultivating more restricted fields than those above referred to, and of a character rather special than local. As examples of these I may mention the Entomological Society in Canada, the Historical Society and the Numismatic Society of Montreal.

Did I suppose that this Society would interfere with the prosperity of such local bodies, I should be slow to favour its establishment. I believe, however, that the contrary effect will be produced. They are sustained by the subscriptions and donations of local members and of the provincial legislatures, while this Society must depend on the Dominion Parliament, from which they draw no aid. They will find abundant scope for their more frequent meetings in the contributions of local labourers, while this will collect and compare these and publish such portions as may be of wider interest. This Society will also furnish means of publication of memoirs too bulky and expensive to appear in local Transactions. There should, however, be a closer association than this. It is probable that nearly all the local societies are already represented among our members by gentlemen who can inform us as to their work and wishes. We should therefore be prepared at once to offer terms of friendly union. For this purpose it would be well to give to each of them an associate membership for its president and one or two of its officers, nominated by itself and approved by our council. Such representatives would be required to report to us for our Transactions the authors and subjects of all their original papers, and would be empowered to transmit to us for publication such papers as might seem deserving of this, and to make suggestions as to any subjects of research which might be developed by local investigation. The details of such association may, I think, readily be arranged, and on terms mutually advantageous, and conducive to the attainment of the objects we all have in view.

It would be a mistake to suppose that this Society should include all our literary and scientific men, or even all those of some local standing. It must consist of selected and representative men who have themselves done original work of at least Canadian celebrity. Beyond this it would have no resting-place short of that of a great popular assemblage whose members should be characterised rather by mere receptivity than by productiveness. In this sense it must be exclusive in its membership, but inclusive in that it offers its benefits to all. It is somewhat surprising, at first sight, and indicative of the crude state of public opinion in such matters, that we sometimes find it stated that a society so small in its membership will prove too select and exclusive for such a country as this; or find the suggestion thrown out that the Society will become a professional one by including the more eminent members of the learned professions. If we compare ourselves with other countries, I rather think the wonder should be that so many names should have been proposed for membership of this Society. Not to mention the strict limitations in this respect placed on such Societies in the mother country and on the continent of Europe, we have a more recent example in the National Academy of Science in the United States. That country is probably nearly as democratic in its social and public institutions as Canada, and its scientific workers are certainly in the proportion of forty to one of ours. Yet the original members of the Academy were limited to fifty, and though subsequently the maximum was raised to one hundred, this number has not yet been attained. Yet public opinion in the United States would not have tolerated a much wider selection, which would have descended to a lower grade of eminence, and so would have lowered the scientific prestige of the country.

Science and literature are at once among the most democratic and the most select of the institutions of society. They throw themselves freely into the struggle of the world, recognize its social grades, submit to the criticism of all, and stand or fall by the vote of the majority; but they absolutely refuse to recognize as entitled to places of importance any but those who have earned their titles for themselves. Thus it happens that the great scientific and literary societies must consist of few members, even in the oldest and most populous countries, while on the other hand their benefits are for all, they diffuse knowledge through the medium of larger and more popular bodies, whose membership implies capacity for receiving information, though not for doing original work, and the younger men of science and literature must be content to earn their admission into the higher rank, but have in the fact that such higher rank is accessible to them, an encouragement to persevere, and in the meantime may have all their worthy productions treated in precisely the same manner with those of their seniors.

Finally, we who have been honoured with the invitation to be the original members of this Society, have a great responsibility and a high duty laid upon us. We owe it to the large and liberal plan conceived by His Excellency the Governor-General to carry out this plan in the most perfect manner possible, and with a regard not to personal, party or class views, but to the great interests of Canada and its reputation before the world. We should approve ourselves first unselfish and zealous literary and scientific men, and next Canadians in that widest sense of the word in which we shall desire, at any personal sacrifice, to promote the best interests of our country, and this in connection with a pure and elevated literature and a true, profound and practical science.

We aspire to a great name. The title of "Royal Society" which, with the consent of Her Gracious Majesty the Queen, we hope to assume, is one dignified in the mother country by a long line of distinguished men who have been fellows of its Royal Society. The name may provoke comparisons not favourable to us; and though we may hope to shelter ourselves from criticism by pleading the relatively new and crude condition of science and literature in this country, we must endeavour, with God's blessing on earnest and united effort, to produce by our cultivation of the almost boundless resources of the territory which has fallen to us as our inheritance, works which shall entitle us, without fear of criticism, to take to ourselves the proud name of the Royal Society of Canada.

Proceedings and Transactions of the Royal Society of Canada, 1 (1882), pp. xi-xvii. (Translation.)

CHAUVEAU, VICE-PRESIDENTIAL ADDRESS

Half a century has not yet passed since the two provinces created by the constitution of 1791 were united, after political events that were then regarded as disastrous; scarcely fifteen years have passed since the federal union of the English colonies of North America, that followed the legislative union of upper Canada and of lower Canada; however, if I undertook to present, in detail, all the advances that have been accomplished in the two periods that I have just indicated, I should scarcely have time to speak of our literary past and of the new institution that we are inaugurating today, and that, we are encouraged to hope, will itself constitute a great step forward, the complement of all the others.

The country is now covered with canals and railways, immense and distant regions have been brought close and given up to colonisation, postal and telegraphic communications have multiplied, mines of every kind have been discovered and exploited, our seaborne services, our industry, our commerce have acquired astonishing proportions, new relations have been established with foreign countries, their capital has been attracted to us, new financial institutions have been created, and our population, in spite of a continual exodus towards the United States, has increased in an almost prodigious fashion: there we have material advance!

The true system of constitutional government, of which we have hitherto had but a vain imitation, has been established; municipal government has improved, and if it is the source of many abuses, it is also the cause of much progress; institutions destined to relieve the miseries of humanity have multiplied, thanks to the initiative of religious communities, of charitable societies and of our governments; the oldest established province has made for itself a code of civil law that is beginning to be a source of envy; we have seen the resolution of questions that were rendered very difficult by the religious and social interests of different sections of the population; finally the sphere of action of our public men has grown, and the federal and local careers open to them . . . do not lack able and devoted followers; there we have political and social progress!

The education of the populace has made real and solid progress everywhere; institutions of higher education have become more numerous and have increased their usefulness; special and scientific institutions have been created; literary associations and journalism have sprung into life and vigour, reviews and literary or scientific journals, in spite of the great difficulties that stood in the way of their success, have become established; new publications boldly replacing those that had died in harness: libraries, museums, popular lectures have multiplied, historical work has acquired great importance; finally, a national literature in each of the two languages, that are in our day what Greek and Latin were in antiquity, has seen the light of day and is now beginning to attract European attention; there we have intellectual progress!

I know that there are shadows in this picture, and if I present it in the best light, it is not because I wish to excuse those who have given a character of marked injustice to the great political evolution that was the point of departure for all this progress, nor do I at all wish to blame the men of my nationality, who made . . . so noble and energetic a resistance to the imperial legislation of 1840. It is thanks to that resistance that they obtained for themselves and for us all, gentlemen, the liberties that we share and that we hold so proudly. Without the subsequent struggle, the two great races that constitute the bulk of the populace of our vast confederation would not have been placed on a footing of equality and would not fraternize at all as they do today.

Moreover, at the most critical moments of our history, there were always English statesmen who understood the role that these two races had to play in this part of the American continent. . . .

No less than five descendants of George III [have visited this country since 1840. One such royal visitor was] the heir presumptive to the crown, who opened the great tubular Victoria Bridge, one of the marvels of America and of the entire world, and who laid the foundation stone of the edifice in which we hold our meetings. May one not believe that the good will of which this great colony has thus been the object is a family tradition . . . ?

The fine arts, under the patronage of Her Royal Highness the Princess

Louise and of His Excellency the Governor-General, have already seen the establishment of an academy whose first exhibitions aroused great hopes; today it is the turn of Science and of Letters.

Science and Letters . . . ; what a world of things in those two words! What they represent is, however, neither so new nor so incomplete in this country as is generally believed. Noble efforts for the culture of the human spirit have been made on the banks of the St. Lawrence for a long, long time. Our history, if we maintain a sense of proportion, is somewhat like that of the Middle Ages, so long ignored or travestied. . . .

Well, since the first foundations established in this country, not only have there been constant efforts to manifest the truths of religion, to establish the practice of the finest virtues that it teaches, the charity to which so many monuments, some of which still exist, were raised, but men have worked with great zeal and activity to transplant the arts and sciences to our land, and to make them flourish here, at the same time that they threw so bright a lustre across the continent of Europe.

It has been established that most of the first colonists could write and read—several were even men endowed with a classical or professional education; we also know that schools were opened in several places, independently from Jesuit institutions, the seminary founded by Mgr. Laval, and the Sulpician Seminary. . . .

The Jesuit college at Quebec, the school of arts founded by Mgr. Laval at Saint Joachim, produced valuable [graduates], some of whom rendered important services to the community. Theses were publicly defended, in imitation of what was done in the old world; governors and intendants assisted and took part in the disputations. These functionaries, like the bishop, were almost always men of letters. Frontenac was a friend of literature, his wife was an intimate in the circle of Mme. de Sévigné. M. de la Galissonnière was a *savant*. Talon was a man of the finest education; M. Dupuy, one of his successors, brought his very considerable library to this country. M. Boucher, governor of Trois-Rivières, wrote a natural history of the country. The missionaries were most often at one and the same time apostles, diplomats, and explorers in the field of science. Father Charlevoix and Father Lafiteau carried out studies in ethnology, and made precious botanical discoveries.

The great explorers did not venture into the vast regions of the west without having the astronomical and geodesic knowledge necessary for their ventures. What is believed to be an instrument for observation, lost by Champlain in his first voyage in the region of the Ottawa, has recently been found. This great man, whom one may call the father of our country, was also a scholar and a vigorous and trustworthy writer. He has left us not only the history of his voyages to Canada, but also a treatise on the art of navigation and a magnificent description of the countries along the Gulf of Mexico, in which his knowledge in the art of drawing and in all branches of natural history is apparent. Besides that, he was the first to conceive the project of uniting the two oceans by a canal across the Isthmus of Panama, a project that one of his compatriots is now carrying out after more than two and a half centuries.

Nicolet, Joliet, Marquette, Gauthier de La Veyrenderie based their discoveries on scientific data. Joliet was a pupil of the Jesuit college, where he defended a thesis that drew attention to him. More than one botanist of this period went into our forests, and Sarrasin and Gauthier had already given their names to indigenous plants, before Kalm, a Swedish pupil of Linnaeus, came to the Château Saint Louis to accept hospitality from M. de la Galissonnière. . . . M. Talon caused the mineral resources and the geography of the country to be studied over a wide region; he therefore needed men of science in his service.

Moreover, in this little world, so cut off during our long winters, stirred by material preoccupations that irresistably imposed themselves, always prey to the emotions of some new war, of some new invasion, it was a marvel to see any science and literature maintain and preserve itself. And yet what charm there is in the *Relations* of the period, what joyous and elegant style, and above all what warmth, what elevation, what profound philosophy in the letters of that mystic, who predicted the greatness of our country, and whom Bossuet called the Saint Theresa of Canada! A taste for beauty, for the ideal, a feeling for nature, that means poetry; the search for truth is philosophy; the study of the world and of its laws is science. These activities are not only encountered in books. . . . All that I mean to say is that there was . . . an intellectual activity in Canada that emerged in a thousand ways but left no written trace, save in a small number of works printed in France and selling today for their weight in gold. Nevertheless civilization has triumphed over barbarism.

Was it not an admirable spectacle to see this small society, concentrated in three small towns, part of it scattered at immense distances, returning with true but almost incredible accounts of all that it had seen and suffered (alas, only too often never returning!); was it not a marvelous spectacle given to the world by that valiant vanguard of civilization, whose role, in certain respects, was the reverse of that of Christian society in the Middle Ages? . . .

For many years the population of the colony was small; the educated class formed a considerable proportion of it, and was, by necessity, intimately mixed with that class less favoured with respect to enlightenment; there was necessarily illumination from one to the other. Missionaries— and every curate in those times was a missionary—the religious orders themselves had only the savages to convert. Wherever they went, they maintained civilization and some degree of education in their constant exchange with even the most distant and most isolated rural population.

Two of the most famous orders were devoted to our country. One of them is famous throughout the world, and Canada provided it with some of the finest pages of its history. Less known than the Jesuits, the work of the Franciscans contributed no less to the great task of civilization. . . .

If our youths were somewhat frivolous in their tastes and habits, as Charlevoix and Kalm charge, it is no less true that centres of light and science existed then as they do today, and one would be quite wrong to believe that the mass of the population was plunged in the thick darkness of ignorance. I admit that after the conquest there was almost a gap. I say it without bitterness, but not without emotion, there was a fairly long period of time in which we were the disinherited of the two nations: our ancient mother country had abandoned us, our new mother country had not yet adopted us. Almost all the educated class, with the exception of the clergy, of some gentlemen and a few lawyers, went back to France; the two religious orders of which I have just spoken were suppressed, all their schools were closed. No more contact with France, no more books. Fortunately the press was not slow in establishing itself here: our first editions, our Canadian incunabula, were school books, prayer books or books of law. They answered the most pressing needs. The periodical press took time in establishing itself; in the beginning, it was of very little help from the literary and political points of view.

Two sites of illumination had remained, however; our seminaries in Quebec and Montreal. Thanks to these two institutions, when constitutional government was established, there was among the French-Canadians, as much as and perhaps more than among those of British origin, a body of men prepared for parliamentary struggles. Panet, Papineau the Elder, Pierre Bédard, de Lotbinière, Taschereau, Blanchet, were our first political glory. Later, Papineau the younger, Vallières, Viger, La Fontaine, Morin and a host of others followed in their footsteps. Politics also gave us our first writers: Bédard and Blanchet in the *Canadien* in 1810—later Morin and Parent. . . .

Science was cultivated in our colleges. M. Bédard, Demers, and several others were its worthy adepts. I shall only mention as a reminder Mr. Wilkie's school, where remarkable men were formed; as well as the Royal Institution and the previously projected university, although these led to no appreciable results. The legislature and the church had established parish schools that were already numerous in 1836, when the government grant was suppressed on legislative advice; finally several new colleges had arisen to help those of Quebec and Montreal. In 1837 there was again a halt in the progress of primary instruction; but the classical or secondary instruction, whose results were indicated by Lord Durham's report, and described as too abundant, continued to expand.

If I now move to institutions like the one that we are inaugurating today, I find that the first attempt of this kind was made in 1809. The Literary Society established at Quebec in that year, took for its motto *Floreamus*

Immemoribus, a well-chosen motto since at that time . . . one could see a forest that extended as far as Hudson's Bay, even from the walls of Quebec. . . .

This first society was not long-lived. It is the same with the first publications, the first reviews, the first associations of this kind, as it is with soldiers who are first in the attack; those who follow and triumph have had to pass over their bodies.

The Literary and Historical Society of Quebec, founded by Lord Dalhousie in 1824, and still in existence today, succeeded the Literary Society of 1809, after rather a long interval. It published numerous memoirs, and some of the most striking men from both national groups are among its active members. After 1848 it had for its rival the *Institut Canadien* of Quebec, which was preferred by educated French-speaking youth.

The Natural History Society, the Historical Society, the Numismatic and Archaeological Society established at Montreal, the Canadian Institute in Toronto, the Geographic Society at Quebec, the French-Canadian Institute at Ottawa, and several other associations of the same kind established in the other provinces of the Federation, and to which the president has just offered a well-deserved tribute, worked and are still working for the propagation of science and the arts.

The task that belongs to such institutions is a hard one in a comparatively new country. It consists of two very different things, the progress of science and the arts, and their popularisation. They necessarily feature little of the academy and much of the lecture-theatre and the public library. As education advances, as literature takes shape and rises to higher regions, as high scientific careers are created and develop, the two functions that I have indicated may separate, and institutions of a more exclusive and elevated character may, with the aid of our governments, become established and prosper.

Have we reached this point? The time for asking this question is past; it was decided by a high and impartial authority which passed a more favorable judgement upon our intellectual and literary movement than we ourselves should have dared to pass.

I have given you a rapid and quite inadequate historical sketch of this movement in the most ancient province in our federation. In recent years it has greatly accelerated everywhere. The great universities of Laval, McGill, Toronto, Lennoxville, Dalhousie, numerous colleges, normal schools, a more complete organisation of public instruction, have spread the taste for science and the arts everywhere. Literary and scientific publications have become numerous, the works of our writers are now known beyond the borders of our country.

For us, descendants of the first colonists, the times have changed greatly since that unlucky time when we were the disinherited of the two nations!

Today, our new mother country grants us enlightened protection, and opens before us the route to prosperity and to a social importance whose limits are hard to assess. On the other hand, our old mother country has remembered us, it behaves most graciously to us, and very advantageous relations are being established between Old and New France, as in the days of Colbert and Talon. . . .

It was thus an opportune moment to convoke this other literary and scientific parliament in the precincts of the Ottawa parliament; our parliament is less clamorous than the one that usually sits here, but its debates, without giving rise to so much passion, will not be entirely without importance and without usefulness.

Men of our two nationalities will meet here, men of all shades of opinion, from all parts of the country. All the sciences will fraternize with one another, and literature and history will join hands with science.

In these days of trial, science has a more difficult mission than ever; its responsibility is also greater. It has been reproached for entering in open war with revealed religion, of sapping every idea of morality by its destructive materialism, indeed of denying both divine activity and human conscience. On the other hand, the powerful physical agents that it has discovered and put in the hands of the vulgar have already given a terrible sanction to these pernicious doctrines; if we do not take care, the moral ruin that it will wreak in men's souls will be followed by terrifying material ruin!

From this point of view, it is a most reassuring guarantee to have at the head of our new institution a man who has fought for so long, and with so much success, for the religious idea in the domain of science, and who, in this respect, has acquired a well-deserved reputation in the United States and Europe. . . .

Proceedings and Transactions of the Royal Society of Canada, 5 (1887), pp. xv–xxii. (Translation.)

THOMAS-ETIENNE HAMEL, PRESIDENTIAL ADDRESS, ROYAL SOCIETY OF CANADA

The principal aim of our society is the encouragement of works of cultivated intelligence, and last year I had occasion to call attention to the immensity of the field open to our researches. . . .

But are educated youths in our country in a position to take part in these serious studies, and to advance science, not simply as a work of the imagination, but as a careful study of the facts, in order to draw rigorous conclusions from them? Alas! I can clearly see obstacles in their path. Allow me to point out three of them, that I would gladly compare with the plagues of Egypt, at least where they concern French Canada. They are journalism, the civil service, and politics. Let me hasten to explain myself.

Far be it for me to think that these three careers are, in themselves, opposed to the development of intelligence; but I mean that, given the conditions under which they are practised in our country, these careers are like tombs where the living forces of our educated youth sicken and die. Undoubtedly there are noble exceptions, but what general rule does not admit of exceptions? That there are some exceptions is a proof of what I am obliged to maintain, because what is now the exception ought to be the general rule. But I realize I owe you some explanations.

Our educated youths are generally poor; and not only is it necessary to live, it is necessary for one to create a position for oneself in order to sustain a family. But the educated young man has a prejudicial view of how he will support himself; he despises manual labour and, once he feels he has attained a certain degree of education, he thinks it beneath him to devote himself to farming, or to blacken his hands in a factory, at least not until he has instantly become the boss, and that without having passed through an apprenticeship. As a consequence, our educated youths wish to live by what is thought of as intellectual work. Because of this, the congestion in the so-called liberal professions has forced a goodly number of its members to seek other additional means of support. But, outside these professions, there remain only those of journalism, governmental positions, and politics—conforming to the common prejudice.

It is not difficult to demonstrate that journalism in this country is not a career that allows one to study seriously. The necessity for quick production allows for only the smallest staff, and results in a considerable and absorbing amount of journalistic work being thrown on to a small number of people. Besides, our journalism exacts very little; it is not fussy about grammar, and, in fact, it demands little in the attainment of knowledge. All that one ever needs to satisfy the urgent needs of the moment is a certain natural facility in jotting down a column or two on any subject whatever, to sustain some argument or other, without even having studied it, and sometimes against one's convictions, . . . Thus . . . journalism in these conditions is unfavourable to the thorough study of true science. It supposes that science has been acquired, but leaves few resources for its acquisition.

If we now pass to the civil service, we shall once again find that such an obstacle to serious studies is present. Parents incapable of ensuring the futures of their families through their own resources have long been greatly preoccupied with obtaining government posts for their children. Government leaders at all levels know how their cause is obstructed by this steeple-chase in which the runners are far more numerous than places for them! In the last few years, they thought they would diminish the number of demands by raising the intellectual qualifications required. But the severity of the mandatory examination, which forces the candidates to undertake great preliminary studies, does not seem to have diminished the number of those who desire to make a career in the civil service. Evidently it is need that is the cause of it. . . .

I have often heard the remark: "But whatever became of such-and-such young fellow, who was so talented and who shone in college? One never hears him spoken of." It has so often been necessary to answer: "Alas! that young fellow, for whom we had such hopes, has buried himself in the tomb of the civil service and, without a miracle, he will not be raised from the grave. . . ."

Let us pass to politics. Oh! Politics! Permit me to despise it, where it concerns our educated youth. Journalism and the civil service are small things in comparison with politics as engines of destruction of the intellectual future of our youth.

Unhappy the young man who, on leaving his collegiate course of studies, distinguishes himself during his clerkship by a certain facility for words, by the talent for improvisation! Once he becomes a lawyer, or notary, or doctor, he will need a great fund of energy and conviction to resist the pressures that will be exerted upon him. The various political parties go and try to corner him; they will clearly show him that his co-operation is absolutely necessary, that the party, and consequently the country, will owe certain triumph to him, etc.

And if the unfortunate fellow allows himself to be led, what will happen?—It seems to me that I can still see young men, whom I knew, whom I tried to persuade by all the arguments inspired by my sincere friendship for them, not to become active in politics until after at least ten years had been given over to the practice of their professions, and the serious study of social questions, . . . it seems to me, I say, that I see them, in the ardour of those fierce fights that are called hotly contested elections and where, consequently, their co-operation is more needed. In a state of continual over-excitement, running from one township to another, occupied with the struggle day and night, obliged, in order to keep going, to mask the weakening of their strength by artificial stimulants, . . . these young men become used to a life of completely external activity. . . .

Another effect of our present political system, paralyzing for the acquisition of science, is the pressure produced on government leaders in their exercise of patronage. All too often it results in not taking merit, science, and aptitude into account when handing out patronage; and positions, no matter how responsible, or at whatever step in the social ladder, are granted under the pressures of irrational influences, to the detriment of the public service.

The examples I could cite, without going beyond my own knowledge! . . .

However, it is not necessary to become discouraged in the presence of all these obstacles, whose remedy will in the long run be furnished by public opinion. It is our task to give a good direction to public opinion, as far as we are able.

Faced with a vast field of studies, let us try to help educated youths throw themselves with ardour in the direction of true science, and not exhaust themselves in vain efforts. Let them persevere in it for a reasonable time. They can be confident that they will end by triumphing over all obstacles and by being worthily repaid.

Transactions of the Manitoba Historical and Scientific Society (1889), 3-6.

PRESIDENT'S INAUGURAL ADDRESS

At a regular monthly meeting of the Manitoba Scientific and Historical society, President C.N. Bell delivered the following inaugural address to the officers and members:

This being the first meeting of the society since the annual election of officers, I take the opportunity afforded me of addressing you on the subject of what should be our lines of work during the coming year. The avowed object of the existence of our society is "to collect and maintain a general library of scientific and popular literature, also to embody, arrange and preserve a library of books, pamphlets, maps, manuscripts, prints, papers, and paintings; a cabinet and museum of minerals, archaeological curiosities and objects generally illustrative of the civil, religious, literary and natural history of the lands and territories north and west of Lake Superior; to rescue from oblivion the memory of the early missionaries, fur traders and settlers of the aforesaid lands and territories, and to obtain and preserve narratives in print, manuscript, or otherwise, of their adventures, labors and observations; to ascertain, record and publish, when necessary, information with regard to the history and present condition of the said regions; and the society may take steps to promote the study of history and science by lectures and otherwise."

The society has not been idle during the years of its existence, having collected a large number of books, pamphlets, papers and manuscripts, with a very fair museum, in addition to publishing thirty-three papers read by members and friends at our monthly meetings. Much has been done, but we have a wide and fertile field of research to labor in, and I desire to suggest some definite lines of work that call for our prompt attention, in the hope that members will be induced to take an interest in them, and afterwards give us the result in the form of papers for publication. . . .

To members who are of a geological turn of mind there is a rich field to explore. Our museum bears evidence of the fact that within a short distance of these rooms there is an abundance of material for the palaeontologist to collect and study. The daily papers frequently contain items of information regarding the revelations made of earth and rock formations in the boring of wells throughout the Province, and with but little labor and inquiry sufficient data can be collected, on this subject, to form the basis of a paper of practical value. Papers on mineralogy dealing with the gold, silver, lead, iron, asbestos and other deposits of the Lake of the Woods and Lake Winnipeg districts; the marble, gypsum, salt and petroleum formations of the region surrounding lakes Manitoba, Winnipegosis, Dauphin and Swan, and the coal beds of the Souris and Saskatchewans will be timely and welcome as agents for drawing attention to the mineral resources of this vast territory.

Meteorology should prove to us a fruitful subject. The Government's records, as published, do not extend to beyond the year 1871, but there are within reach journals and books which give more or less information covering odd portions of the present century. Important data of the areas affected by early frosts, with the degree of severity experienced in the elevated plateaus and lowlands, the wooded and plain districts, the land adjoining lakes and rivers and on the open prairie, the light and heavy soils, etc., might be collected and turned to practical benefit in the future. The cause of extraordinary high water in our rivers, resulting at times in floods, is to be investigated and placed on record, while the periodical rise and fall of the water in lakes like Manitoba and the Lake of the Woods, if described and the causes explained, would be instructive.

The action of frost on soils in Manitoba has received some attention from members of this society, but extended observations must be made, during a series of years, to obtain sufficient material to found any trustworthy conclusions upon.

The botanical field is almost a virgin one. As the dairy interests of this Province are becoming a prominent feature of the country's sources of wealth, a carefully prepared paper on the native grasses and the vegetable growths injurious to animals, or inimical to the production of high grade dairy produce, is one that would be well received and bring the society's practical usefulness to the notice of the public.

Some of our medical members have probably given attention to the medicinal preparations in use among the Indians, as well as their forms of application. The result of their observations and enquiries in this direction would prove a fitting subject for a paper. That we have Seneca root, collected by Indians and exported from the Province to the United States, of a value yearly of several thousands of dollars, is well known. A list of our trees and shrubs, the ascertained limits of their growth, and details of the wild fruits, indigenous to Manitoba would be of service. We have enthusiastic mushroom hunters as members. Will one of them not give us a paper on the edible fungi of the Province? So little is known of the value of this form of food, in a country producing spontaneously such a large number of varieties, that really good services would be rendered by the publication of an article plainly describing the forms and their usual places of growth. Wheat, barley and oats have been raised in the Red River Valley during the greater part of the present century. Where did the seed come from, and did resowing year after year result in any distinct change of quality or yielding power? It is a well known fact that farmers in this province, living even but a score or two of miles apart, find that an exchange of seed now results in an improved yield and quality. As being a matter of extraordinary consequence to a great grain growing district like ours, this question, in all its bearings, is one well worthy of investigation by the society.

Entomology will offer many inducements to members who take an interest in it—and there is much need of their services being utilized. The grasshopper visitations to this region, while few in number, can be traced, and if full information regarding them is collected and recorded it may be of service. While singularily free so far from the ravages of insect pests, we can scarcely hope to escape them altogether in the future, and all items of interest relating to them should be gathered and kept on file. Largely dependent, as the people of Manitoba are, on the successful growth of grain, anything liable to affect the growing crops should be closely observed and studied.

The habits and resorts of the wild animals of the Northwest ever forms an instructive and interesting line of study, and it is to be observed that the advance of civilization is surely driving away animals abundant but a few years ago. The grizzly bear figures as a leading feature in the "prairie tales" of twenty-five years back. The traders' journals written at posts on the Saskatchewans contain almost daily entries of encounters with these enormous animals, but passengers in the comfortable sleeping cars now glide smoothly along through the depleted hunting grounds without obtaining a sight of one of these monarchs of the prairie. Sixteen years ago, in the country now traversed by the Canadian Pacific Railway west of Medicine Hat, standing on a hill, I saw, as far as the eye could reach, great moving masses of buffalo, and now the only remnant left of these countless tens of thousands, are less than two hundred animals existing in the wooded fastnesses of the Athabasca and Peace Rivers. The curious emaciation in the ranks of the northern rabbits every eight or ten years offers a subject for investigation. The fur returns of the Hudson Bay Company, during a wide range of years, reveals the fact that every ten years the catch of lynx drops off suddenly from hundreds of thousands to one-tenth of that number, and as the lynx depends for food mainly on the rabbit supply, there is evidently some connection between the periodical disappearance of both. The effect produced on the prairie soil by the excavating and deep plowing propensities and power of the badger, gopher and mole, indulged in for a long succession of years, is known to have been considerable, and has been the subject of remark and discussion. With ample opportunities for observation, is it too much to ask a member of our Natural History section to study up this matter and give us the result of his research? None of the society's publications deal with the fishes found in the lakes and rivers of the Northwest. In Lake Winnipeg alone there are said to be 17 varieties, and it appears that there is some confusion in the names locally given to families of the pike and white-fish varieties. A difference of opinion now exists as to whether our lake fisheries are being injured by over fishing, and a very important point to be taken into consideration, in getting at the truth of the situation, is the extent to which fish migrate from one part of the lake to another. and the cause of their appearance and disappearance in certain portions of the lakes in different years. I have no doubt but that very important evidence of these points may be obtained by inquiry.

Northwest birds have already received attention at the hands of the society, but there is plenty of room for ornithologically inclined students to work in yet.

So far as I know, reptilian life in this country has not been investigated and reported on. Who will undertake to identify the frog members of the orchestra that make melody during the summer evenings. The well-known "snake hole" at Stony Mountain, with notes of the annual autumn gathering of the reptiles, would alone furnish ample material for an interesting paper.

Our museum is sadly lacking in collections of fresh water and land shells. A paper on the varieties found in this Province might enlist the services of a corps of collectors.

While the National museum of the United States has a very extensive and valuable collection of birds' eggs, gathered in Manitoba and the Territories by correspondents of the Smithsonian Institute, our museum is without one identified specimen—who of our members will undertake to make a collection for our own museum?

The moundbuilders' remains in Manitoba have by no means all been examined or studied. There is a circular embankment a few miles north of Gretna that awaits exploration, and on the Assiniboine near Virden several mounds are situated which should receive attention, before the plowshare destroys them. Excavations made at Point Douglas and Fort Rouge have revealed the remains of animals and charred wood at from six to twelve feet below the surface of the ground, and as time passes other "finds" of like nature will undoubtedly occur. Every effort should be made to secure reliable information concerning them, to place on record in our archives for future use. . . .

In my opinion, in view of the fact that the society receives some slight amount from the Provincial Government, one of the prime objects to be held in view by our members is the securing and distribution of information of a practical character which will be beneficial to the people of the whole Province.

CHARLES N. BELL,
President.

B. GENESIS, GEOLOGY, AND EVOLUTION

J.W. Dawson, *The Chain of Life in Geological Time* [n.d.], pp. 259-69.

We may, then, I think, make the following affirmations:—

(1) The existence of life and organisation on the earth is not eternal, or even coeval with the beginning of the physical universe, but may possibly date from Laurentian or immediately pre-Laurentian times.

(2) The introduction of new species of animals and plants has been a continuous process, not necessarily in the sense of derivation of one species from another, but in the higher sense of the continued operation of the cause or causes which introduced life at first. This, as already stated I take to be the true theological or Scriptural as well as scientific idea of what we ordinarily and somewhat loosely term creation.

(3) Though thus continuous, the process has not been uniform; but periods of rapid production of species have alternated with others in which many disappeared and few were introduced. This may have been an effect of physical cycles reacting on the progress of life.

(4) Species, like individuals, have greater energy and vitality in their younger stages, and rapidly assume all their varietal forms, and extend themselves as widely as external circumstances will permit. Like individuals, also, they have their periods of old age and decay, though the life of some species has been of enormous duration in comparison with that of others; the difference appearing to be connected with degrees of adaptation to different conditions of life.

(5) Many allied species, constituting groups of animals and plants, have made their appearance at once in various parts of the earth, and these groups have obeyed the same laws with the individual and the species in culminating rapidly, and then slowly diminishing, though a large group once introduced has rarely disappeared altogether.

(6) Groups of species, as genera and orders, do not usually begin with their highest or lowest forms, but with intermediate and generalised types, and they show a capacity for both elevation and degradation in their subsequent history.

SIR JOHN WILLIAM DAWSON

(7) The history of life presents a progress from the lower to the higher, and from the simpler to the more complex, and from the more generalised to the more specialised. In this progress new types are introduced, and take the place of the older ones, which sink to a relatively subordinate place, and become thus degraded. But the physical and organic changes have been so correlated and adjusted that life has not only always maintained its existence, but has been enabled to assume more complex forms, and that older forms have been made to prepare the way for newer, so that there has been on the whole a steady elevation culminating in man himself. Elevation and specialisation have, however, been secured at the expense of vital energy and range of adaptation, until the new element of a rational and inventive nature was introduced in the case of man.

(8) In regard to the larger and more distinct types, we cannot find evidence that they have, in their introduction, been preceded by similar forms connecting them with previous groups; but there is reason to believe that many supposed representative species in successive formations are really only races or varieties.

(9) In so far as we can trace their history, specific types are permanent in their characters from their introduction to their extinction, and their earlier varietal forms are similar to their later ones.

(10) Palaeontology furnishes no direct evidence, perhaps never can furnish any, as to the actual transformation of one species into another, or as to the actual circumstances of creation of a species, but the drift of its testimony is to show that species come in *per saltum*, rather than by any slow and gradual process.

(11) The origin and history of life cannot, any more than the origin and determination of matter and force, be explained on purely material grounds, but involve the consideration of power referable to the unseen and spiritual world.

Different minds may state these principles in different ways, but I believe that in so far as palaeontology is concerned in substance they must hold good, at least as steps to higher truths. And now I may be permitted to add that we should be thankful that it is given to us to deal with so great questions, and that in doing so deep humility, earnest seeking for truth, patient collection of all facts, self-denying abstinence from hasty generalisations, forbearance and generous estimation with regard to our fellow-labourers, and reliance on that Divine Spirit which has breathed into us our intelligent life, and is the source of all true wisdom, are the qualities which best become us.

As we have traced onward the succession of life, reference has been made here and there to the defects of those bold theories of descent with modification which are held forth in our time as the true bond of the links of the chain of life. It must have been apparent that these theories, however specious when placed in connection with a limited induction of facts selected for the purpose of illustrating them, are very far from affording a satisfactory solution of all difficulties. They cannot perhaps be expected to take us back to the origin of living beings; but they also fail to explain why so vast numbers of highly organised species struggle into existence simultaneously in one age and disappear in another, why no continuous chain of succession in time can be found gradually blending species into each other, and why in the natural succession of things degradation under the influence of external conditions and final extinction seem to be laws of organic existence. It is useless here to appeal to the imperfection of the record or to the movements or migrations of species. The record is now in many important parts too complete, and the simultaneousness of the entrance of the faunas and floras too certainly established, while the moving of species from place to place only evades the difficulty. The truth is that such hypotheses are at present premature, and that we require to have larger collections of facts. Independently of this, however, it would seem that from a philosophical point of view all theories of evolution, as at present applied to life, are fundamentally defective in being too partial in their character; and this applies more particularly to those which are "monstic" [gnostic] or "agnostic," and thus endeavour to dispense with a Creative Will behind nature. . . .

. . . These hypotheses are partial, inasmuch as they fail to account for the vastly varied and correlated interdependencies of natural things and forces, and for the unity of plan which pervades the whole. These can be explained only by taking into the account another element from without. Even when it professes to admit the existence of a God, the evolutionist reasoning of our day limits itself practically to the physical or visible universe, and leaves entirely out of sight the power of the unseen and spiritual, as if this were something with which science has nothing to do, but which belongs only to imagination or sentiment. So much has this been the case that when recently a few physicists and naturalists have turned to this aspect of the subject, they have seemed to be teaching new and startling truths, though only reviving some of the oldest and most permanent ideas of our race. From the dawn of human thought it has been the conclusion alike of philosophers, theologians, and the common sense of mankind, that the seen can be explained only by reference to the unseen, and that any merely physical theory of the world is necessarily partial. This, too, is the position of our sacred Scriptures, and is broadly stated in their opening verse; and indeed it lies alike at the basis of all true religion and all sound philosophy, for it must necessarily be that "the things that are seen are temporal, the things that are unseen, eternal." With reference to the primal aggregation of energy in the visible universe, with reference to the introduction of life, with reference to the soul of man, with reference to the heavenly gifts of genius and prophecy, with reference to the introduction of the Saviour Himself into the world, and with reference to the

spiritual gifts and graces of God's people, all these spring not from sporadic acts of intervention, but from the continuous action of God and the
unseen world; and this, we must never forget, is the true ideal of creation
in Scripture and in sound theology. Only in such exceptional and little influential philosophies as that of Democritus, and in the speculations of a
few men carried off their balance by the brilliant physical discoveries of
our age, has this necessarily partial and imperfect view been adopted.
Never indeed was its imperfection more clear than in the light of modern
science.

Geology, by tracing back all present things to their origin, was the first
science to establish on a basis of observed facts the necessity of a beginning and end of the world. But even physical science now teaches us that
the visible universe is a vast machine for the dissipation of energy; that
the processes going on in it must have had a beginning in time, and that
all things tend to a final and helpless equilibrium. This necessity implies
an unseen power, an invisible universe, in which the visible universe must
have originated, and to which its energy is ever returning. The hiatus between the seen and the unseen may be bridged over by the conceptions of
atomic vortices of force, and by the universal and continuous ether; but
whether or not, it has bec[o]me clear that the conception of the unseen as
existing has become necessary to our belief in the possible existence of the
physical universe itself, even without taking life into the account.

It is in the domain of life, however, that this necessity becomes most
apparent; and it is in the plant that we first clearly perceive a visible testimony to that unseen which is the counterpart of the seen. Life in the
plant opposes the outward rush of force in our system, arrests a part of
it on its way, fixes it as potential energy, and thus, forming a mere eddy,
so to speak, in the process of dissipation of energy, it accumulates that on
which animal life and man himself may subsist and assert for a time
supremacy over the seen and temporal on behalf of the unseen and eternal.
I say, for a time, because life is, in the visible universe, as at present constituted, but a temporary exception, introduced from that unseen world
where it is no longer the exception but the eternal rule. In a still higher
sense, then, than that in which matter and force testify to a Creator, organisation and life, whether in the plant, the animal, or man, bear the
same testimony, and exist as outposts put forth in the succession of ages
from that higher heaven that surrounds the visible universe. In them, as
in dead matter, Almighty power is no doubt conditioned by law, yet they
bear more distinctly upon them the impress of their Maker, and while all
explanations of the physical universe which refuse to recognise its spiritual and unseen origin must necessarily be partial and in the end incomprehensible, this destiny falls more quickly and surely on the attempt to
account for life and its succession on merely materialistic principles.

Here, however, we must remember that creation, as maintained against such materialistic evolution, whether by theology, philosophy, or Holy Scripture, is necessarily a continuous, nay, an eternal influence, not an intervention of disconnected acts. It is the true continuity, which includes and binds together all other continuity.

It is here that natural science meets with theology, not as an antagonist, but as a friend and ally in its time of greatest need; and I must here record my belief that neither men of science nor theologians have a right to separate what God in Holy Scripture has joined together, or to build up a wall between nature and religion, and write upon it "no thoroughfare." The science that does this must be impotent to explain nature and without hold on the higher sentiments of man. The theology that does this must sink into mere superstition.

In the light of all these considerations, whether bearing on our knowledge or our ignorance, a higher and deeper question presents itself, namely, that as to the relation of nature and of man to a Personal Creator. To this it seems to me that the study of the succession of life yields no uncertain reply. Call the progress of life an evolution if you will; trace it back to primeval Protozoa, or to a congeries of atoms: still the truth remains that nothing can be evolved out of these primitive materials except what they originally contained. Now we find in the existence of man, and in the tendency of the scheme of nature towards his introduction, evidence that at least all that is involved in the reasoning and moral nature of man must have existed potentially before atoms began to shape themselves into crystals, or into organic forms. Nay, more than this is implied, for we do not know that man and what he has hitherto been and done constitute the ultimate perfection of nature, and we must suspect that something much more than what we see in man must be required for the origination of the chain of life. What does this prove, in any sense in which human reason can understand it? Nothing less, it seems to me, than that doctrine of the Almighty Divine Logos, or Creative Reason, as the cause of all things, asserted in our sacred Scriptures, and held in one form or another by all the greatest thinkers who have attempted to deal with the question of origins. Falling back on this great truth, whether presented to us in the simple "God said" of Genesis, or in the more definite form of the New Testament, "The Word was with God, and the Word was God," we find ourselves in the presence of a Divine plan pervading all the ages of the earth's history and culminating in man, who presents for the first time the image and likeness of the Divine Maker; and this forms the true nexus of all the separate chains of life. Had man never existed, such reasoning might have been speculative merely, but the existence of man, taken in connection with the progress of the plan which has terminated in his advent, proves the existence of God.

J.W. Dawson, *The Origin of the World According to Revelation and Science*, 6th ed. (London, 1893), pp. 343-59.

The parallelism of . . . conclusions of careful inductive inquiry into the structure of the earth's crust, with the results . . . from revelation, may be summed up under the following heads:

    1. Scripture and Science both testify to the great fact that there was a beginning—a time when none of all the parts of the fabric of the universe existed; when the Self-Existent was the sole occupant of space. The Scriptures announce in plain terms this great truth, and thereby rise at once high above atheism, pantheism, and materialism, and lay a broad and sure foundation for a pure and spiritual theology. . . .

Yet science, and especially geological science, can bear witness to this great truth. The materialist, reasoning on the fancied stability of natural things, and their inscription within invariable laws, concludes that matter must be eternal. No, replies the geologist, certainly not in its present form. This is but of recent origin, and was preceded by other arrangements. Every existing species can be traced back to a time when it was not; so can the existing continents, mountains, and seas. Under our processes of investigation the present melts away like a dream, and we are landed on the shores of past and unknown worlds. But I read, says the objector, that you can see "no evidence of a beginning, no prospect of an end." It is true, answers geology; but, in so saying, it is not intended that the present state of things had not an ascertained beginning, but that there has been a great and, so far as we know, unlimited series of changes carried on under the guidance of intelligence. These changes we have traced back very far, without being able to say that we have reached the first. We can trace back man and his contemporaries to their origin, and we can reach the points at which still older dynasties of life began to exist. Knowing, then, that all these had a beginning, we infer that if others preceded them they also had a beginning. But, says another objector, is not the present the child of the past? Are not all the creatures that inhabit the earth the lineal descendants of creatures of past periods, or may not the whole be parts of one continual succession, under the operation of an eternal law of development? No, answers geology, species are immutable, except within narrow limits, and do not pass into each other, in tracing them toward their origin. On the contrary, they appear at once in their most perfect state, and continue unchanged till they are forced off the stage of existence to give place to other creatures. The origin of species is a mystery, and belongs to no natural law that has yet been established. Thus, then, stands the case at present. Scripture asserts a beginning and a creation. Science admits these, as far as the objects with which it is conversant extend, and the notions of eternal succession and spontaneous development, discountenanced both by theology and science, are obliged to take refuge in those misty regions where modern philosophical skepticism consorts with the shades of departed heathenism.

2. Both records exhibit the progressive character of creation, and in much the same aspect. . . . This geological order of animal life, it is scarcely necessary to add, agrees perfectly with that sketched by Moses, in which the lower types are completed at once, and the progress is wholly in the higher.

In the inspired narrative we have already noticed some peculiarities, as, for instance, the early appearance of a highly developed flora, and the special mention of great reptiles in the work of the fifth day, which correspond with the significant fact that high types of structure appeared at the very introduction of each new group of organized beings—a fact which, more than any other in geology, shows that, in the organic department, elevation has always been a strictly *creative* work, and that there is in the constitution of animal species no innate tendency to elevation, but that on the contrary we should rather suspect a tendency to degeneracy and ultimate disappearance, requiring that the fiat of the Creator should after a time go out again to "renew the face of the earth." In the natural as in the moral world, the only law of progress is the will and the power of God. In one sense, however, progress in the organic world has been dependent on, though not caused by, progress in the inorganic. We see in geology many grounds for believing that each new tribe of animals or plants was introduced just as the earth became fitted for it; and even in the present world we see that regions composed of the more ancient rocks, and not modified by subsequent disturbances, present few of the means of support for man and the higher animals; while those districts in which various revolutions of the earth have accumulated fertile soils or deposited useful minerals are the chief seats of civilization and population. In like manner we know that those regions which the Bible informs us were the cradle of the human race and the seats of the oldest nations are geologically among the most recent parts of the existing continents, and were no doubt selected by the Creator partly on that account for the birthplace of man. We thus find that the Bible and the geologists are agreed not only as to the fact and order of progress, but also as to its manner and use.

3. Both records agree in affirming that since the beginning there has been but one great system of nature. . . .

4. The periods into which geology divides the history of the earth are different from those of Scripture, yet when properly understood there is a marked correspondence. Geology refers only to the fifth and sixth days of

creation, or, at most, to these with parts of the fourth and seventh, and it divides this portion of the work into several eras, founded on alternations of rock formations and changes in organic remains. The nature of geological evidence renders it probable that many apparently well-marked breaks in the chain may result merely from deficiency in the preserved remains; and consequently that what appear to the geologist to be very distinct periods may in reality run together. The only natural divisions that Scripture teaches us to look for are those between the fifth and sixth days, and those which within these days mark the introduction of new animal forms, as, for instance, the great reptiles of the fifth day. We have already seen that the beginning of the fifth day can be referred almost with certainty to the Palaeozoic period. The beginning of the sixth day may with nearly equal certainty be referred to that of the Tertiary era. The introduction of great reptiles and birds in the fifth day synchronizes and corresponds with the beginning of the Mesozoic period; and that of man at the close of the sixth day with the commencement of the Modern era in geology. These four great coincidences are so much more than we could have expected, in records so very different in their nature and origin, that we need not pause to search for others of a more obscure character. It may be well to introduce here a tabular view of this correspondence between the geological and Biblical periods, extending it as far as either record can carry us, and thus giving a complete general view of the origin and history of the world as deduced from revelation and science.

5. In both records the ocean gives birth to the first dry land, and it is the sea that is first inhabited, yet both lead at least to the suspicion that a state of igneous fluidity preceded the primitive universal ocean. In Scripture the original prevalence of the ocean is distinctly stated, and all geologists are agreed that in the early fossiliferous periods the sea must have prevailed much more extensively than at present. Scripture also expressly states that the waters were the birthplace of the earliest animals, and geology has as yet discovered in the whole Silurian series no terrest[r]ial animal, though marine creatures are extremely abundant; and though air-breathing creatures are found in the later Palaeozoic, they are, with the exception of insects, of that semi-amphibious character which is proper to alluvial flats and the deltas of rivers. . . .

6. Both records concur in maintaining what is usually termed the doctrine of existing causes in geology. Scripture and geology alike show that since the beginning of the fifth day, or Palaeozoic period, the inorganic world has continued under the dominion of the same causes that now regulate its changes and processes. The sacred narrative gives no hint of any creative interposition in this department after the fourth day; and

Biblical Aeons	*Periods Deduced from Scientific Considerations*
The Beginning.	Creation of Matter.
First Day.—Earth mantled by the Vaporous Deep—Production of Light.	Condensation of Planetary, Bodies from a nebulous mass—Hypothesis of original incandescence.
Second Day.—Earth covered by the Waters—Formation of the Atmosphere.	Primitive Universal Ocean, and establishment of Atmospheric equilibrium.
Third Day.—Emergence of Dry Land—Introduction of Vegetation.	Elevation of the land which furnished the materials of the oldest rocks —Eozoic Period of Geology?
Fourth Day.—Completion of the arrangements of the Solar System.	Metamorphism of Eozoic rocks and disturbances preceding the Cambrian epoch—Present arrangement of Seasons—Dominion of "Existing Causes" begins.
Fifth Day.—Invertebrates and Fishes, and afterward great Reptiles and Birds created.	Palaeozoic Period—Reign of Invertebrates and Fishes. Mesozoic Period—Reign of Reptiles.
Sixth Day.—Introduction of Mammals—Creation of Man and Edenic Group of Animals	Tertiary Period—Reign of Mammals. Post-Tertiary—Existing Mammals and Man.
Seventh Day.—Cessation of Work of Creation—Fall and Redemption of Man.	Period of Human History.
Eighth Day.—New Heavens and Earth to succeed the Human Epoch—"The Rest (Sabbath) that remains to the People of God."*	

---

*Heb. iv., 9; 2 Peter iii., 13.

geology assures us that all the rocks with which it is acquainted have been produced by the same causes that are now throwing down detritus in the bottom of the waters, or bringing up volcanic products from the interior of the earth. This grand generalization, therefore, first worked out in modern times by Sir Charles Lyell, from a laborious collection of the changes occurring in the present state of the world, was, as a doctrine of divine revelation, announced more than three thousand years ago by the Hebrew lawgiver; not for scientific purposes, but as a part of the theology of the Hebrew monotheism.

7. Both records agree in assuring us that death prevailed in the world ever since animals were introduced. . . .

8. In the department of "final causes", as they have been termed, Scripture and geology unite in affording large and interesting views. They illustrate the procedure of the All-wise Creator during a long succession of ages, and thus enable us to see the effects of any of his laws, not only at one time, but in far distant periods. To reject the consideration of this peculiarity of geological science would be the extremest folly, and would involve at once a misinterpretation of the geologic record and a denial of the agency of an intelligent Designer as revealed in Scripture, and indicated by the succession of beings. Many of the past changes of the earth acquire their full significance only when taken in connection with the present wants of the earth's inhabitants; and along the whole course of the geological history the creatures that we meet with are equally rich in the evidences of nice adaptation to circumstances and wonderful contrivances for special ends, with their modern representatives. As an example of the former, how wonderful is the connection of the great vegetable accumulations of the ancient coal swamps, and the bands and nodules of iron-stone which were separated from the ferruginous sands or clays in their vicinity by the action of this very vegetable matter, with the whole fabric of modern civilization, and especially with the prosperity of that race which, in our time, stands in the front of the world's progress. . . .

The Bible's . . . harmony with natural science, as far as the latter can ascend, gives to the Word of God a pre-eminent claim on the attention of the naturalist. The Bible, unlike every other system of religious doctrine, fears no investigation or discussion. It courts these. "While science," says a modern divine,* "is fatal to superstition, it is fortification to a Scriptural faith. The Bible is the bravest of books. Coming from God, and conscious of nothing but God's truth, it awaits the progress of knowledge with calm security. It watches the antiquary ransacking among classic ruins, and re-

---

*Hamilton

joices in every medal he discovers and every inscription he deciphers; for from that rusty coin or corroded marble it expects nothing but confirmations of its own veracity. In the unlocking of an Egyptian hieroglyphic or the unearthing of some implement it hails the resurrection of so many witnesses; and with sparkling elation it follows the botanist as he scales Mount Lebanon, or the zoologist as he makes acquaintance with the beasts of the Syrian desert; or the traveller as he stumbles on a long-lost Petra or Nineveh or Babylon. And from the march of time it fears no evil, but calmly abides the fulfilment of those prophecies and the forthcoming of those events with whose predicted story inspiration has already inscribed its page. It is not light but darkness which the Bible deprecates; and if men of piety were also men of science, and if men of science were to search the Scriptures, there would be more faith in the earth, and also more philosophy."

C. THE DIFFUSION OF USEFUL KNOWLEDGE—SCIENCE AND AGRICULTURE

N. Aubin, *La Chimie Agricole mise à la portée de tout le monde* (Québec, 1847), Introduction. (Translation.)

The question now being discussed, that will, I hope, be resolved entirely to the nation's advantage, is whether Canada must become an exclusively agricultural or manufacturing country. Her still virgin mineral riches, her geographical position, her numerous navigable waterways, her countless waterfalls that provide a sum of motive power superior to all those that are operating today at great expense throughout the rest of the world, indubitably invite her children to throw themselves into the arena of industry; but one must not let oneself be blind to the advantages that cannot accrue without great agricultural prosperity. It is evident that, *without cheap bread*, there can be no production of inexpensive goods. The proof of this proposition is eloquently furnished by the recent measures that England owes to its parliament, enlightened by the clear-sighted spirit of RICHARD COBDEN.

It was in the general interest, and in order to maintain Britain's industrial supremacy, that English economists brought in agricultural produce from abroad, to the immediate prejudice of the private interests of the home farmers. The mother country's intelligent but costly experience should be an example to us, and should warn us that we must henceforth aim solely at increasing the produce of our soil, while doing our utmost to diminish both manual labour and foolhardy trials that are in no way founded on the strict teachings of science.

The health of a country grows by the spread of useful knowledge of all kinds, and only when our agriculturists become as enlightened about the theory of their art as they are able and industrious in its practice will they be able to call to their aid and fearlessly risk capital that is now seeking other destinations.

A manufacturing population can only prosper when surrounded by an equally prosperous agricultural population. Fields and work-shops must mutually and turn about lend one another arms, instruments, healthy and abundant food. Without agriculture, industry becomes a misfortune for the country that yields itself up to it; it is a source of moral and physical weakness, profitable only to those who would exploit it.

Agriculturist and Canadian Journal (1848), 50.

AGRICULTURAL COLLEGE, &C.

We are determined not to lose sight of this great desideratum—a suitable Institution for the Education of Farmers' Sons in *their* PROFESSION. There is nothing like doing things in the "nick of time." Our legislature is now assembled. One great Public Institution, the UNIVERSITY, is expected to be finally settled, and while the subject is under consideration, we shall not omit to put in our claim, nor, as the farmer's organ, fail to remind our Representatives, that as a *class*, the agriculturists of Canada demand some adequate provision for the establishment and support of an Educational Institution for *their* advancement. We think they have every right to claim a direct interest as a class, in the University revenue. The agriculturists of Canada, and agriculturists every where, are the "*first* class," in the noblest and best sense. The Merchants, Mechanics, Priests, Lawyers, Artists, Literati, &c., &c., are all non-producers—mere hangers-on, dependants of the husbandman. He can do without them, they cannot live without him. If you wish to see genuine virtue, true patriotism, unostentatious benevo-lence, sterling honesty and practical piety, go among the cultivators of the soil. Look not for these rarities in the crowded city; they will not vegetate in the tainted atmosphere that surrounds the haunts of busy, plotting rivalry, priestly intrigue, scheming political selfishness, legal trickery, and reckless commercial gambling. Even in a country so young as Canada, with a changing, heterogeneous population, the truth of this contrast becomes every day plainer to the view. The sturdy yeomen are the true conserva-tives of society. They are the substratum—the foundation of the social fabric—and if that be defective, the whole building will tumble in ruins. It has been so in all past time, in all other countries: it is so in ours. Why then should it be thought unnecessary to afford every facility for the acquisition of knowledge by farmers? Is the *common* school good enough for them? Is it because they are as a class, compared with others, virtuous, patriotic, benevolent, honest, &c., that they need not also be intelligent? Must we give 8 or 10,000 pounds a-year for the support of Professors, with their philosophic apparatus and appliances, and scholarships, and prizes, and low charges for tuition, in order that a few citizens may educate their sons for the *learned* professions, while not £1000 is given to support an Institution for teaching sound principles to those who are intended for a profession, not "learned," it may be, but vitally necessary; a profession, in the pursuit of which, the lights of modern science, the discoveries and improvements of modern times, are absolutely essential to complete and certain success? Must the "arts" be encouraged, while the "nursing mother of all the arts" is left to shift for herself?

But we leave this broad, general view for the present, and come to the £ *s. d.* aspect of the question. We said the farmers of Canada have a *right* to claim a direct interest in the University revenue. It may be answered, so they have, and will have the privilege of sending their sons to be educated within its walls, on the same terms as others. But they dont want the *kind* of learning to be obtained there. It is not suited to them, unless they wish to become Lawyers or Doctors—either of which will probably be the very worst use that farmers can put their sons to. No, the farmers of this country, as such, will be practically excluded from the benefits of the University, unless a portion of its funds be appropriated for the support of an Agricultural School. So far as we can learn, the estate, *i. e.* the *lots of land*, and their proceeds, which the hard work of farmers, in clearing away the bush, and in making roads around these "reserves," has rendered valuable, and to the benefits of which they are therefore pre-eminently entitled, is quite sufficient to sustain a University properly conducted, suitable to the present wants of the country, and a School of Agriculture, on a respectable scale, besides. We admit, that in one view, it will be quite indifferent, as to the public source from which the appropriation come, provided it *do* come, and be sufficient in amount. But we know that the public funds are already mortgaged almost beyond redemption, and we are not going to expose ourselves to the objection that we show what is wanted, without showing that it *can* be granted. We wont take the answer—"we have no funds." We point to the available means, and we assert our *right* to participate. If other provision be made, we shall not complain, though as economists, and using our right as constituents to judge of the proper course to be pursued by our representatives in dealing with public property, we are of opinion, that the University funds are the legitimate means for such a purpose.

Agriculturist and Canadian Journal (1848), 74-5.

CAN A FARMER LEARN TO ANALYZE HIS SOIL?

To the Editors of the Agriculturist

GENTLEMEN,—

As a farmer in this region of country, I, like many others, have laboured under great disadvantages from the want of a scientific knowledge in my profession. A want, which I am necessarily compelled to believe, and to acknowledge as painfully vexatious in its circumstances, and ruinous in its consequences, compared with the profitable and satisfactory results which proceed from a judicious and scientifically systematic mode of cultivation. Numerous statements have been made by learned agriculturists, both theoretical and practical, of the great advantages that may be, and that are derived from pursuing a systematic course of farming, conducted on principles in accordance with laws physically organized, but from these advantages, we in this isolated portion of the country, have hitherto been precluded from a participation in, partly from our contracted means, and partly from a want of inclination. With a climate by no means uncongenial to the successful cultivation of a great variety of fruits and vegetables, and with a soil abundantly productive, yet, notwithstanding our most assiduous industry, combined with all the skill which long experience has made us master of, our crops sometimes prove a complete—often a partial failure. The question is often asked, how can we account for these things? Conjecture is set afloat to investigate into the cause of the failure of this crop and into the failure of that crop, and every cause assigned but the true one. Many of my neighbor farmers who suffer in common with myself, are beginning to awake to a true sense of their position, and are beginning earnestly to enquire after the light of science, and the most eligible means of its attainment, which if once attained would no doubt form a new epoch in the system of farming in this part of Canada West. But how are we to attain this? Would you or some of your correspondents be kind enough to inform us whether, if we were to purchase scientific books, we could progress so far as to be able to analyze the different soils, or whether, if we were to unite, we could procure the services of a professional man for some two or three months through the course of the season, to lecture for us, and teach us to analyze the different properties of the soils? Would you be kind enough to inform us also, what apparatus would be necessary, and what would be the expense?

Yours respectfully,

THOMAS BOYLE.

Sandy Point, near Amherstburg, }
    20th March, 1848.   }

In reply to Mr. Boyle, we may state, that a complete and exact analysis of a soil requires the skill of a practical analytical chemist. Indeed, to be able to state accurately all the ingredients contained in a given soil, with the relative proportions of each, supposes the possession of the highest skill; the process is extremely nice and difficult, and often requires weeks to complete it. But there are three or four substances that are essential in all soils, to the healthy growth of the ordinary crops. Potash, lime, magnesia, &c. If a soil, for instance, be entirely, or nearly deficient in lime, it would not raise a good crop of clover or lucerne, while it would produce rye grass in abundance, and even a fair crop of wheat, if not deficient in other substances which these latter require. Now it does not demand a practised chemist to tell whether a soil be deficient in lime. An intelligent man may soon learn enough chemistry to ascertain such a point. A mere tyro may apply tests, and satisfy himself whether lime be present in large or very small quantities. And the same may be said of potash or soda. Again, wheat and other grain crops require a large amount of phosphoric acid. If the soil be deficient in this, it will be impossible to raise good crops. A partial knowledge of chemistry will enable a person to determine that question, and the same partial knowledge will point out to him the substances or manures which contain this ingredient, viz: bone dust, guano, &c. The nature and quantity of the various ingredients contained in the different manures are, of course, explained by chemists, and may be ascertained by reference to their works.

We mean therefore to say, that a sufficient amount of chemical knowledge may be obtained from study and observation, by an intelligent farmer, without resorting to the laboratory of the practical chemist to enable him to form a tolerably correct judgment of the nature of his soil, and whether any of the substances absolutely essential to the growth of crops are in excess or deficient. He will probably be able to learn enough in this way for all the practical purposes of cultivation. Professor Johnston's Catechism of Agricultural Chemistry, a little book of 74 pages, contains a rich mine of information, almost sufficient in itself for the purposes mentioned. His lectures on chemistry and geology are more full and scientific in their character. The works of Liebig, and numerous other writers on scientific agriculture, may be studied by the intelligent farmer with great advantage.

We know of no person in this country qualified to give lectures, and illustrate them by experiments, except Mr. Buckland. Professor Crofts, (of the University), is, we believe, a first-rate analytical chemist, and if we were anxious to know the constituent elements of our soil, we should prefer sending a portion to him to be analyzed with scientific accuracy. Mr. Buckland studied under Professor Johnston, the best agricultural chemist of his day, and might, we dare say, be engaged to deliver a course of lectures to any club or society that would pay a reasonable remuneration.

An apparatus sufficient to perform all the experiments mentioned in

Johnston's Catechism, may be had in Albany, N. Y., for about four dollars. We are not aware that any thing additional would be required, except perhaps a few tests, to make an *unprofessional* examination of a piece of soil.

Canadian Agriculturist (1850), 147.

AGRICULTURE AND CHEMISTRY

We should be sorry to insinuate the slightest doubt of the ability of chemistry to assist the farmer in the practical details of his daily vocation; something has already been done in advancing the art of agriculture, and much more, we believe, remains to be accomplished, by invoking the aid of science; but to suppose that any practicable educational system will ever convert the farmers of a country (that is, such as pursue their calling for a living) into expert analytical chemists, appears to us perfectly wild and visionary. Many that speak and write upon these matters seem to have no definite conception of the time and patience, the deep and accurate knowledge, with the habit of delicate manipulation, which are required in every satisfactory analysis of organic compounds. The farmers must remain content to leave this business in the hands of those to whom such matters properly belong. An imperfect analysis is worse than useless, for any purpose, either practical or theoretical.

Canadian Agriculturist (1851), 123-4.

GEOLOGICAL SURVEY OF CANADA

Blackfriars Mills, London, C.W. 8th April,
1851

To the Editor of the Canadian Agriculturist

Dear Sir,

Numbers of your subscribers in the London and Western Districts, have, up to this time, been anxiously expecting (through the medium of your useful columns, or some other source,) the report of our Canadian geologist, through this section of the Province, about two years ago. When at London, that gentleman visited one of our best farmers, Mr. Christopher Walker, in the 12th Concession, and also the flats of our beautiful river Thames, and took specimens of the soil, with a promise that we should be furnished with a statement of their several qualities and requisites in the spring following. That time has passed and another at hand, yet nothing

THE

CANADIAN AGRICULTURIST;

A MONTHLY JOURNAL

OF

Agriculture, Horticulture, Mechanics and General Science, Domestic Economy, &c.

profit of the earth is for all; the King himself is served by the field."—Eccles. v. 9.

GEORGE BUCKLAND,
WILLIAM McDOUGALL,

EDITORS AND
PROPRIETORS.

VOL. I.

TORONTO:

PRINTED BY ROWSELL & THOMPSON.

1849.

has appeared to satisfy the curiosity excited. Do, if you please, in your next publication, as our agricultural instrument, find out something relative to the important question,—What this part of the country requires, especially the river flats, to realize better crops?

I am, Dear Sir,

Your obedient servant,

ROGER SMITH

P.S.—Our fall wheat looks admirable in this section of the country and around Goderich.

R.S.

[Anxious to meet the wishes of our subscribers in the London and Western Districts, as referred to by our correspondent, we subjoin from Mr. Logan's Geological Report for 1849-'50, such portions as bear upon the objects of the enquiry. It is to be regretted that these valuable reports are not better known in the remoter districts of the country. Some of our readers may not be aware of the fact, that T.S. Hunt, Esq., is the chemist and mineralogist to the Geological Survey, and consequently the analysis of soils falls within his department. In the fall of 1849, Mr. Hunt collected forty specimens of soils from different parts of Upper and Lower Canada, the results of such as he had been enabled to analyze are appended to the before mentioned report, and the remainder we should suppose will appear in the next. Few of our readers have any conception of the time and attention required in making a correct and thorough analysis of a soil; several of the operations are of the most delicate nature, requiring the best modern apparatus, with the minutest attention and most advanced knowledge of the manipulation. Many analys[e]s of soils that have been made and published, are next to useless, for any practical purpose; indeed, not a few of them, from the imperfect manner in which they have been performed, will positively mislead. From Mr. Hunt's well known industry and attainments as an analytical chemist, the public may place the utmost confidence in the accuracy of his results.]

Canadian Agriculturist (1850), 265-6.

PROFESSORSHIP OF AGRICULTURE IN THE UNIVERSITY OF TORONTO

Farmers of Canada; you who have subdued the forest and by your indomitable industry dotted over the country with comfortable homesteads and thriving settlements; you who are the main source of our wealth and prosperity, will soon, we are happy to hear, be directly represented in the highest Educational Institution in this Province. As intimated in our last, the authorities of the University are contemplating a plan for filling the chair of Agriculture, and converting a portion of the University grounds into an *Experimental Farm*; and by what fell from the Chancellor, at the meeting of Convocation the other day, we are led to expect, that these valuable objects will be speedily carried into effect. The press, as was to be expected, has favourably noticed the movement and we should hope, for the character and good of the country, that no factious opposition would be offered to its progress and successful termination. We are not in possession of the full particulars of the scheme which is now before the Senate; but we understand that it is proposed to place the whole of the University grounds, consisting of about 180 acres, under the superintendence of the Professor of Agriculture; whose duties are not to be confined to the delivery of class lectures on mere scientific, or even practical subjects, connected with husbandry and rural affairs; but a sufficient portion of those grounds that are already cleared of timber, comprising some 70 or 80 acres are to be devoted to the purposes of experimental and practical farming. The land is to be given up for a term of years free of charge to the Professor, subject to the control of the Board of Agriculture; —an important, and what it is hoped will prove, a most useful instrumentality, that is about being organised under the provisions of an act of the Legislature, passed last session. We learn upon good authority that the Government will recommend to Parliament a sufficient grant of money for carrying out the important objects of the Board, and for sustaining with increased vigor the Provincial Association. The country should distinctly understand that this is no mere political movement for party purposes; its object is purely patriotic, and it should enlist the sympathies and support of all who sincerely desire their country's welfare. We live in an age and are now placed in circumstances, which imperatively demand, that the improvement of agriculture, the main source of our wealth, should receive the earnest attention and support of the Legislature, irrespective of what party may control the helm. Many other countries, our near and enterprising neighbors in particular are prosecuting this object with an earnestness and intelligence that cannot fail of success; and amidst the increasing competition of the civilized world with the markets of the mother country equally thrown open to all, it will not do for Canadians to fold their arms in listlessness, and to stand still, while the rest of the world is rapidly moving onwards. Not a moment ought to be lost. We must be up and doing; bringing willingly to our aid whatever science or experience can suggest for increasing the fertility of our fields, and for developing those great natural sources of wealth and enjoyment, which a bountiful Providence has placed within our reach.

The recognition of the claims of agriculture by the University, cannot fail to render that important Institution more popular and useful, in a country where four-fifths of its inhabitants are engaged in the cultivation of the soil. The social *status* of our farmers will become elevated, by associating the Science and practice of their pursuits, with a liberal course of academical learning; while existing systems of farm practice, will be necessarily improved, by imparting to the young, sound, practical knowledge, and the results of carefully conducted experiments. We will now proceed to give our views of this matter a little more in detail.

In the first place, whatever is attempted should bear upon the face of it, the stamp of *practical utility*. The lectures of the Professor on the theory and practice of his art, ought to be fully illustrated, not only by diagrams, specimens and models, but especially by frequent reference to the daily operations of the farm. The merely pointing out the application of some of the laws and doctrines of chemistry, geology, animal and vegetable physiology, &c., to the pursuits of the farmer, however interesting and suggestive as many of these undoubtedly are, would be quite a different thing from the practical teaching of agriculture as an *art*. The principle on which a Professorship of agriculture should be founded in the present day, according to our notion, is that of *Practice with Science*.

Canadian Agriculturist (1858), 170-3.

EXAMINATION PAPER IN AGRICULTURE, UNIVERSITY COLLEGE,
TORONTO, SESSION 1857-8

Thinking it probable that a large number of our readers have no very definite notion of the nature and extent of the examination of students in Agriculture in our Provincial Collegiate Institute, it may not be devoid of use and interest to publish entire an examination paper belonging to this department. The one subjoined was written by Mr. J. E. Farewell, of Oshawa, in the County of Ontario, and obtained the first prize. . . . [He] had been more or less engaged in practical farming, and availed [himself] of several other courses of lectures in the College, besides Agriculture, during the winter season. Young men can enter the College as occasional students, without being subjected to any preliminary examination, and may attend such courses of lectures only as meet their more immediate wants. Agricultural students usually take in addition to the history, science, and practice of Agriculture, Chemistry, Geology and Mineralogy, Natural History, including Botany and Meteorology, History, and English Language and Literature. Youths intended for the business of farming can annually go through such, or, if need be, a more extended course of study, and not be absent from their farms during the busiest and most important seasons of the year. All this can be done for a comparatively small expense; but little exceeding that for board and lodging. It is proper to state that the terminal examinations in the College are conducted on the principle of written answers to a series of questions, to each of which is assigned a numerical value; the students being strictly prohibited from any intercourse with books, notes, or each other, during the period of examination. The agricultural examination occupied two sittings, of two hours each. We give of course the paper as it was written, with only an occasional verbal correction.

Question 1.—Define Agriculture as a *science* and an *art.*

How can a knowledge of its *Theory* and *Practice* be best acquired?

Answer.—1. Agriculture as a *science*, treats of the principles or laws which govern the operations of converting the inert matter of earth, air and water, into vegetable productions for the support of animal life.

2. As an *art* Agriculture treats of the *application* of these principles to practical purposes. The former gives the rules of the operations, and the reasons for them. The latter applies the rules advanced by science.

The best mode of acquiring a thorough knowledge of the science and practice of Agriculture, authors of high repute seem somewhat to differ.

Stephens—a good authority in practice—says, this can be best done by living with a farmer, who is a good practical man, and who has in his house an instructor in the theory or science, daily pointing out on the farm the practical application.

On the other hand, Professor Norton is in favour of the plan adopted in University College—attending lectures on the theory and practice as given by a Professor, and afterwards reducing their principles to practice on the farm during the active season of sowing, growth and maturity. To such as have had some experience on a farm, this seems the most suitable plan; or instead o[f] this, attending some Agricultural School with a suitable farm attached, and learning both theory and practice at the same time.

Question 2.—Mention those branches of physical science which have relations to Agriculture;—with illustrations.

Answer.—The various departments of Physics, or Natural Philosophy relating to Agriculture are—

1. Motion, the moving powers, their nature, laws and operation, the effects of machinery. Mechanics.

2. The weight, pressure and equilibrium of fluids. Hydrostatics.

3. The motion of fluids in pipes or otherwise, and their capability and value as moving powers. Hydraulics.

4. The action of light on vegetation. Optics.

5. The nature, laws and effects of heat.

6. The laws of electricity, and other meterological phenomena.

7. The nature of air as regards its properties of weight, temperature, motion, &c., and the signs which foretell these movements. Pneumatics.

8. Chemistry, explaining the nature and composition of all bodies, and the laws of their combination.

9. Botany, treating of the structure, uses and classification of plants.

Also including vegetable physiology, explaining their functions, diseases, &c.

10. Zoology, relating to the structure and classification of the animal kingdom, with which is connected comparative anatomy and physiology of the domesticated animals of the farm.

11. Geology, explaining the structure and arrangements of rocks, their origin and diffusion, with the decomposition whereby soils are formed.

Question 4.—How is matter divided? Define and illustrate *elementary, compound, organic* and *inorganic* substances? What are soil, plants and animals composed of?

Answer.—Matter exists in the following states, viz:—solid, liquid, gaseous and vesicular. A familiar example is water, which by being exposed to a low temperature, becomes a solid, (ice) which again is liquified by heat, and by still further heat is converted into an invisible vapour (steam).

An elementary substance is matter that cannot be reduced to a simpler form; *e. g.,* iron, oxygen, sulphur, &c. A compound body is that which is made up of two or more elementary substances; *e. g.,* oxide of iron or rust, consisting of oxygen and iron, sulphate of potassa, composed of sulphur and potassium, &c.

Organic substances are the result of life, in the vegetable or animal, and by heat become decomposed and converted into invisible gases; *e. g.,* carbonic acid, oxygen, hydrogen, &c.

Whereas inorganic bodies do not consume by heat, were never the seat of any sort of life, being purely mineral; *e. g.,* iron, silica or sand, iodine, manganese, &c.

Soils are generally composed of a number of different substances, the principle being clay, sand, and lime, potash, soda, magnesia, manganese, &c., are more or less found in connection with organic substances in all fertile land.

Plants consist mainly of carbon, oxygen, and hydrogen, with small portions of nitrogen, combined with the several substances mentioned in soils.

Animals consist of the same organic elements constituting plants, but with a much larger proportion of nitrogen, and a very great amount of the phosphate of lime in the bones, so valuable as a manure.

Question 5.—State the composition and uses of *atmospheric air and water,* and their relations to vegetable and animal life.

Answer.—Atmospheric air mainly consists of two gases, nitrogen and oxygen; about 79 parts of the former and 21 of the latter in every 100 of common air. There are also diffused through the atmosphere small quantities of carbonic acid gas, ammonia, and some aqueous vapour.

Water consists of a chemical combination of oxygen and hydrogen, in the proportion of 8 of the former with 1 of the latter. This is pure rain water, but the waters of springs, rivers, &c., have in them a number of other ingredients, as lime, soda, &c., in varying proportions.

James Fletcher, 'Practical Entomology', *Proc. & Trans. Roy. Soc. Can.*, S2, 1 (1895), 3-7.

Undoubtedly a noticeable feature of the last decade has been the general recognition of the value of science,—that is, accurate knowledge—in carrying on all the ordinary occupations of life. The foolish ideas that science is a sort of wonderland, not to be entered except by a favoured few, or that science as a study must not be popularized for fear of degrading it, are now only held by the ignorant or those who are unwilling to learn. Science is, merely, accurate knowledge in all branches of study, and the popularizing of science means only the rendering of such knowledge so simple or accessible as to be available to all who wish to learn.

Strenuous efforts are now being made by the leading thinkers in all lines of study, to give their investigations a practical application to the every-day affairs of life. In no direction have these efforts been attended with so much success as in what are called the natural sciences. Recent developments in the application of electricity may well be said to have revolutionized the whole systems of transportation and communication, as well as the lighting of cities and individual buildings, and offer a most attractive field for discussion by any scientific body. Perhaps I need not crave your indulgence for drawing your attention to a few instances showing how the agriculture of to-day is benefited by the investigations of scientific workers. The chemist is now thoroughly recognized by the agricultural classes as the magician who can unlock to them hidden secrets as to the true value of various crops; can tell them which are the best to grow for stock, to provide food, or as fertilizers of the soil; can tell them, after analysis, what constituents of a soil are lacking, and advise them as to the most economical way of replacing the required elements. The skill of the practical botanist is now being chiefly directed to the examination of parasitic fungi, with the object of devising suitable remedies for those species which attack cultivated vegetation, or of propagating the parasitic forms which destroy insect life. In the closely allied branches of horticulture and agriculture, numberless experiments are being carried on daily with the object of discovering and originating by selection and hybridization new varieties of flowers, fruits, grains and vegetables; the best methods of propagation and cultivation, the best treatment of the soil and the most effectual and economical fertilizers. Here also, in passing, reference may be made to the care, improvement and treatment in health and disease of all farm stock. All of the above work comes under the head of scientific agriculture—provided that it is done accurately—and is of incalculable value to the country at large. In no branch of natural science, however, I believe, have such important results been obtained, when gauged by their effect upon the revenue of the country, as in that branch of zoology which treats of insects and their depredations on the crops of the orchard, the garden, and the farm. . . .

As stated above, great advance has been made on this continent, which is certainly due to the practical tendency of the majority of the people of North America to recognize what is useful when laid plainly before them; but also to the excellent nature of the work which has been done. It would be entirely out of the question to mention even the names of the many eminent economic entomologists of the United States, and only such will be cited now as are necessary in speaking of the few instances which I propose to lay before you, to illustrate a few of the heads of subjects embraced in a consideration of the practical use of the study of entomology.

In Canada, outside of the work done by the Dominion Entomologist's Department at Ottawa, founded in 1884, the development of economic entomology has been most intimately connected with that of the Entomological Society of Ontario, which issued its first report on injurious insects in 1870. Since that year these valuable reports have appeared regularly, and have supplied the farmers of Ontario with much information of incalculable value, which has certainly saved the country a great deal of unnecessary loss.

Special mention must be made of the classic work on "Insects Injurious to Fruits," by our fellow-member, Prof. W. Saunders. He and the Rev. Dr. Bethune, the present able editor of the "Canadian Entomologist," were both original members of the Entomological Society of Ontario, and have borne the brunt of the work of building up that thriving and useful organization which is now presided over by my esteemed colleague and co-worker, Mr. W. H. Harrington, who is now so well known in scientific circles all the world over for the care, accuracy and thoroughness of his work. The late Abbé Provancher published much and did excellent work, but it was chiefly of a scientific and descriptive nature.

In the curriculum of the Ontario Agricultural College at Guelph, practical entomology is included, and Prof. J. H. Panton not only lectures on this subject to the students, but has also published several timely bulletins on injurious insects for the use of farmers. Besides the above, there have appeared some useful articles by Prof. A. H. Mackay in the "Educational Monthly" of St. John, N.B., and an excellent article on the Flour Moth by Dr. Bryce of Toronto. Useful notes on the occurrence of injurious insects have also appeared in the report of the statistician of the Department of Agriculture of British Columbia. In 1894 an Inspector of Fruit-pests was appointed for British Columbia in the person of Mr. R. M. Palmer, and doubtless good results will follow this step.

A fact which should never be lost sight of, in considering the progress and bearings of practical entomology in Canada, is the enormous advantages we enjoy from having so near to us the United States, with its army of trained entomologists and other officials at Washington and at the state colleges and agricultural experiment stations. Most of the insect enemies which are injurious in Canada occur also in some of the States of the Union, and it is mutually advantageous to entomologists here and in the United States to be able to study together, under slightly different circumstances, any new pests which may occur. In addition to this, of course, many of the insects which appear in Canada have already been studied out carefully in the United States before they reach our borders, and we have the benefit of all the experience of our neighbours to guide and help us in counteracting their attacks. In the case of some insects practical and effective remedies have been discovered before the insects invaded our borders. Notable instances of such insects are found in the Cattle Horn-fly, the Pear-leaf Blister-mite, and the Pear-tree Psylla.

William Saunders, 'Agricultural Progress', *Proc. & Trans. Roy. Soc. Can.*, S3, 1 (1907), Appendix A, xli-xliv.

Twenty-three years ago farming in Canada was in a very depressed condition and in 1884 the House of Commons appointed a Select Committee to enquire into this subject and to suggest the best means of developing and encouraging the agricultural interests of this country. Careful investigation led to the conclusion that the general lack of success was not due to any fault of the soil or climate, nor to want of industry among the farmers, but to defective farming from want of skill and knowledge in all branches of this work, and up to this time, no provision had been made by the Government to remedy this. There is probably no industry engaging the attention of mankind that requires more skill and general information to conduct successfully than farming. Competition in food products is keen throughout the civilized world, and the farmer must turn to practical account every advantage within his reach to improve the quality of his products and to lessen the cost of their production if he is to improve his position.

The Committee recommended that the Government should establish experimental farms, where experiments might be carried on in all branches of agriculture and horticulture and that the results of this work should be published from time to time and distributed freely among the farmers of the Dominion.

The recommendations of the Committee were favourably received and early in 1886 an act was passed authorizing the Government to establish a central experimental farm and four branch farms. The central farm was to be located near Ottawa and the branch farms in different parts of the Dominion, one in the Maritime Provinces, one in Manitoba, one in the Northwest Territories and one in British Columbia.

In choosing these sites an effort was made to have them fairly representative in soil and climate of the larger settled areas in the provinces or territories in which they were placed. In the arrangement of the work such experiments as were most likely to be beneficial to the larger number of settlers were in each case among the first to engage the attention of the officers in charge.

Twenty years have passed since this work was begun and during that time agriculture in Canada has made unprecedented advancement. Investigations and experimental researches have been conducted in almost every line bearing on agriculture and horticulture and a multitude of important facts have been accumulated and given to farmers throughout the Dominion in reports and bulletins. The principles which underlie successful crop-growing have been frequently dealt with and demonstrated. The importance of maintaining the fertility of the land, adopting a judicious rotation of crops, following the best methods of preparing the land, early sowing, choosing the best and most productive varieties and the selection of plump and well-matured seed, all these have been shown to be essential to success.

Through the experimental farms early ripening sorts of grain have been brought from many countries wherever they could be found. While none of those tried have been found equal in quality to the best sorts already cultivated here, the new importations have given early ripening strains, which, by skilful crossing and selecting, have already produced excellent results. Several of the newer varieties of wheat ripen from two to three weeks earlier than some of the well known sorts in cultivation, thus opening up a prospect of considerably extending the wheat area in the Canadian Northwest. Distinct gains have also been made by crossing and selection in other classes of cereals. Varieties of grasses, suited to the needs and conditions prevailing in the Northwest have been experimented with and distributed for test, whereby dairying and stock raising are now becoming easier to conduct and more remunerative. New apples also have been produced by crossing very hardy forms of Siberian crab apples with varieties of apples grown in Eastern Canada. These cross-bred sorts have proved quite hardy at several hundred different points at varying altitudes, and are succeeding in those parts of the Northwest country where ordinary apples are too tender to be successfully grown.

Other lines of original research, chemical, botanical and entomological, have also been followed with great assiduity, while the other branches of agricultural and horticultural work have been carried on with similar enthusiasm, and special bulletins on many important subjects have been published and widely distributed.

DR WILLIAM SAUNDERS

The backward condition of agriculture in Canada, which was so pronounced twenty years ago, has given place to one of constant progress and advancement, and, instead of a lack of skill and knowledge among the farmers of this country, I think it may now be safely said that Canadian farmers, on an average, are as well informed and more generally progressive than those of any other country in the world.

The Experimental Farms have been one of the important factors in the educative work of this country and the Government is now wisely adding to the opportunity of Canadian farmers to gain knowledge by increasing the number of these institutions. Two experimental stations have recently been established in Alberta, one in the southern part of the province at Lethbridge, to study the various problems connected with irrigation and dry farming, and one further north at Lacombe to carry on experiments in general farming suitable for that district. Experiments are also in progress under Government direction in the Peace River country and the Yukon. It is expected that other sub-stations will shortly be established on Prince Edward Island, Vancouver Island and in northern Saskatchewan. These will no doubt be followed by others so that eventually these experimental institutions will be sufficiently numerous to meet the needs of our various climates.

The reputation of Canada as an advanced agricultural country stands high, and other nations are earnestly interesting themselves in the fine agricultural products for which Canada is now noted. When the National Miller's Association of Great Britain began their efforts to improve the quality of the wheat grown in the Mother Country, application was made for the best wheats obtainable here and although varieties were obtained by them from many other countries, none have yet been found superior to the best of those sent from Canada. Many other lands have also sought for samples of the agricultural products of this country for trial. Among the British Colonies many different sorts have been sent to Australia, South Africa and Newfoundland. India has applied for some of the best products for test in that country especially in the higher altitudes in the mountain districts. Requests have recently come from Thibet for food materials likely to be grown with success in the high plains of that country at altitudes ranging from 12,000 to 16,000 feet. In response to requests from the Russian Department of Agriculture many varieties of wheat, barley and maize have been supplied which are being tested in different parts of that Empire. Even from Egypt the great granary of early times requests for Canadian grain have been received and the varieties sent are now being tested at Khartoum and along the Nile. To Japan also many different sorts have been forwarded for trial, and quite recently a number of different varieties have been sent for test in that part of the Saghalien Islands which reverted to Japan as a result of the late war. Similar requests have also been recently responded to from Italy and from Mexico. Canada has won an enviable reputation as a country of vast agricultural resources, and the published records of her progress have many interested readers in all countries where intelligent agriculture is practised. Immigrants are flocking to our shores in large and increasing numbers and millions of acres of virgin land are being brought under crop. The mass of surplus food products available for export, shows every year a marked increase while as yet the area of land under cultivation is relatively small. What these exports will amount to in the near future, when the country becomes well settled, and the acreage of crop much larger, no one can accurately foretell. Enough, however, is known to warrant the statement that Canada will shortly become one of the greatest food-exporting countries of the world.

Canadian Agriculturist (1861), 170.

THE ART OF AGRICULTURE

A great deal has been written and said about the science and art of agriculture, but for practical guidance the whole thing is in a nut shell. It consists in these two rules—make the land rich, and keep the weeds down. If any person who tries to raise any plant will follow these two rules he will succeed, and if he does not follow them he will not succeed.

D. WAR AND SCIENCE

R. Lemieux, Resolution, *Proc. & Trans. Roy. Soc. Can.*, S3, 4 (1915), xliii-xliv.

The Honourable Rodolphe Lemieux introduced the following resolution:—
    Whereas this is the first general meeting of The Royal Society of Canada since the breaking out of the war in Europe;
    Whereas the said war is one involving several of the great powers, and especially the mother countries of the two main stocks of people making up this Dominion;
    Whereas Canada is nobly performing her duty in this emergency in training, equipping and sending forth thousands upon thousands of her hardy sons to fight the battles of peace, freedom and humanity on the blood-soaked battlefields of France and Belgium;
    Whereas the type of kultur or civilization which Germany and her ally

Austria, with the help of Turkey, have undertaken to impose by force on the rest of the universe is supremely repugnant to those higher ideals of justice and liberty which it has been the especial burden of the British mother country and its galaxy of self-governing commonwealths to establish;

Whereas the barbarous means resorted to by our foes to attain their sinister ends have entailed the ruthless destruction of many of the finest monuments of architecture, treasures of art and institutions of learning, particularly in Belgium and the north of France;

And whereas The Royal Society of Canada, as the premier Canadian institution representative of the interests of Art, Science and Literature, is bound to manifest its deepest concern in the preservation of such monuments, treasures and institutions; and whereas The Royal Society of Canada glories in having some of its members who answered the call of England against the German wild agression; . . .

Be it Resolved:

That we members of The Royal Society of Canada, duly assembled for our Annual meeting in Ottawa, do strongly voice our loathing of the atrocities and depredations thus committed by our foes, and that we do solemnly enter and make public our desire to join our unanimous protest with those which have emanated from similar institutions the world over against the perpetration of such crimes.

Moved by Hon. Rodolphe Lemieux (President of Section I), seconded by Hon. Mr. Justice Longley (Past-President of Section II), that the resolution be adopted and the motion was unanimously and enthusiastically carried.

'The Honorary Advisory Council for Scientific and Industrial Research in Canada', *Proc. & Trans. Roy. Soc. Can.*, S3, 11 (1917), xvi-xxi.

One of the most striking results of the great war is the sudden awakening of the English speaking world to the importance of scientific and industrial research, and the realization by Governments of the necessity of applying scientific research to the whole range of problems which present themselves in both war and peace.

With the declaration of war the supplies of several classes of products for which Great Britain had come to rely almost exclusively upon Germany—in the manufacture of which that country has gradually secured a practical monopoly—were suddenly cut off. Some of these, such as dye stuffs, optical glass, etc., were of vital importance to certain of Great Britain's industries, which were very seriously threatened by the impossibility of securing adequate supplies of these necessary materials. Some of these materials were even needed for the manufacture of arms and munitions of war, and the necessity of making Great Britain independent of foreign countries not only for the requirement of industry but also for the essentials of national defence was thus made clear.

The Government of Great Britain, having been brought to a realization of these facts, appointed in July, 1915, a committee of the Imperial Privy Council for Scientific and Industrial Research, with an advisory council composed of eight men distinguished in the world of science and industry "for the development of scientific and industrial research" applicable to the problems of war and the development of the industries of peace that follow the war.

The Government of Australia thereupon established "a Commonwealth institute of science and industry" along similar lines. New Zealand and India have also expressed a desire to co-operate with the Imperial Government in every possible way.

If, after the war, the industries and manufactures of the Dominion are to develop and expand in the face of the very rigorous competition which will grow up under the conditions which will follow the declaration of peace, it is necessary that our industrial and manufacturing operations shall be carried on with much more efficiency than has, as a general rule, characterized them in the past.

Taking a leaf out of our enemies' book, and for the purpose of making a first step toward the closer application of science to the industrial development of the Dominion, the Government of Canada on June 6th, 1915, appointed a Committee of the Canadian Privy Council consisting of the Right Honourable the Minister of Trade and Commerce (Chairman), the Honourable the Ministers of the Interior, Agriculture, Mines, Inland Revenue and Labour, to devise and carry out measures to promote and assist scientific and industrial research with a view to the fuller development of Canadian industries and production in order that during and after the present war they may be in a position to supply all Canadian needs and to extend Canadian trade abroad.

Under this committee of the Privy Council there was constituted on the 29th of November, an Honorary Advisory Council for Scientific and Industrial Research, ...

This Advisory Council, by direction of the Chairman of the Committee of the Privy Council, has been charged with the following duties:

(a) To ascertain and tabulate the various agencies in Canada which are now carrying on scientific and industrial research in the universities and colleges, in the various laboratories of the Government, in business organizations and industries, in scientific associations or by private or associated investigators.

(b) To note and schedule the lines of research or investigation that are being pursued by each such agency, their facilities and equipment therefor, the possibilities of extension and expansion, and particularly to ascertain the scientific man power available for research and the necessity of adding thereto.

(c) To co-ordinate these agencies so as to prevent overlapping of effort, to induce co-operation and team work, and to bring up a community of interest, knowledge and mutual helpfulness between each other.

(d) To make themeslves acquainted with the problems of a technical and scientific nature that are met with by our productive and industrial interests, and to bring them into contact with the proper research agencies for solving these problems, and thus link up the resources of science with the labor and capital employed in production so as to bring about the best possible economic results.

(e) To make a scientific study of our common unused resources, the waste and by-products of our farms, forests, fisheries and industries, with a view to their utilization in new or subsidiary processes of manufacture and thus contributing to the wealth and employment of our people.

(f) To study the ways and means by which the present small number of competent and trained research men can be added to from the students and graduates of science in our universities and colleges, and to bring about in the common interest a more complete co-operation between the industrial and productive interests of the country and the teaching centres and forces of science and research.

(g) To inform and stimulate the public mind in regard to the importance and utility of applying the results of scientific and industrial research to the processes of production by means of addresses to business and industrial bodies, by the publication of bulletins and monographs, and such other methods as may seem advisable.

In pursuance of the work with which it has been charged, the Council, in order to develop in Canada a body of men who have been thoroughly trained in science and its application to industry, such as that which has aided so greatly in the industrial development of Germany in recent years, has recommended to the Government the establishment of twenty or more studentships and fellowships in Canadian universities and technical schools, to be given to men who have completed their regular course of study and have displayed a special aptitude for scientific research.

These will enable such men to pursue a course of advanced work for a further period and thus acquire a practical training in the methods and conduct of research. Arrangements are also contemplated whereby students will be placed in one or other of the great manufacturing establishments of the Dominion, where they will continue their training under the conditions of actual commercial practice.

For the purpose of making a complete census or inventory of all work in scientific and industrial research which is being carried on in the Dominion at the present time by all the agencies now at work, and also for the purpose of ascertaining the various lines and directions in which the application of research was most necessary and might be made most fruitful in the development of our industries and manufactures, the Council in the spring of 1917 issued questionnaires to all the Universities, Government Departments, Technical Societies, as well as to all Canadian Manufacturers, asking under definite heads for specific information on the various subjects which come within the purview of the Council. In the distribution and in the collection of proper returns from these questionnaires the Council has received the active, energetic and sympathetic assistance of the various Technical Societies of the Dominion, as well as of the Canadian Manufacturers' Association.

The Council has also enlisted the close co-operation of all the Government Departments, both Federal and Provincial, for the purpose of correlating and rendering more easily accessible the wealth of information concerning the Natural Resources of the Dominion which lie stored in the Government Archives and Reports.

In addition to this broad and general work which looks toward the establishment of a substantial basis for the further development of the industries of the Dominion in the immediate future, the Council has examined carefully a large number of specific projects which have been submitted to it, and has approved of certain of these which appear to give promise of valuable results.

They have decided to recommend that two of these projects be at once taken up and work be started upon them immediately. The first has for its object the provision of an adequate supply of good fuel for the Western Plains, more especially in the Provinces of Saskatchewan and Manitoba. There are in the former Province large supplies of lignite. This is an inferior fuel possessing a relatively low heating power and which, furthermore, will not stand shipment and storage. It is, therefore, of comparatively little value for domestic or manufacturing purposes. The Council, however, believes that by a special treatment there may be produced from this lignite two grades of high class briquetted fuel, one similar to anthracite or hard coal in character, and the other resembling soft coal in general character, and at the same time certain valuable by-products may be secured. The Department of Mines and the Commission of Conservation have already carried out a good deal of investigation in connection with this problem, and the former Department is now making some further studies for the Council. If they give satisfactory results the Council will advise that a plant to turn out this high-grade fuel on a commercial scale be erected, and the possibility of producing this fuel at a cost considerably lower than that at which coal from the United States is now laid down in Manitoba and Saskatchewan be demonstrated on a large scale, the coal being actually placed on the market. With an abundant supply of good cheap fuel the conditions of life on the great plains in winter will be much improved.

The other project has to do with the preservation of the forests of eastern Canada. These, contrary to the opinion which prevails generally, are not inexhaustible. They have already been seriously depleted and are rapidly deteriorating in character. In most of the leading countries of Europe the forests, whether owned by the Government or by private interests, have, by the application of modern scientific knowledge, been immensely improved in character, and, instead of being plundered and then abandoned, have been converted into assets of enormous national value, and which year by year yield large revenues to the Government, or to their private owners, which are as regular and as continuous as those from any other gilt-edged investment, the forest all the time being maintained with its capital unimpaired.

Different methods of forest management have been adopted in different parts of Europe to secure this splendid result. The Canadian forests present special problems of their own. The Council has recommended that the necessary means be provided in order to enable the Forestry Branch of the Department of the Interior to carry out certain investigations for the purpose of ascertaining which of these methods can best be applied to the Canadian forests with a view to stopping the destruction which now threatens them, and of making these forests a great and permanent source of wealth to the people of the Dominion.

Many other projects and many additional lines of work are under consideration by the Council, but these require further examination before the Council is in a position to decide what action should be taken with reference to them.

Proceedings & Transactions of the Royal Society of Canada, S3, 12 (1918), 1-6.

THE WAR AND SCIENCE
By Dr. A. Stanley Mackenzie, f.r.s.c.
(Delivered Tuesday, May 21, 1918)

I know that the general subject upon which I have chosen to address you is one about which much has been said and written of late; but, notwithstanding this fact, and the further fact that I cannot hope to throw new light on the matter, I have felt justified in my temerity by the impression which has been steadily growing upon me that today Science has fallen upon the most momentous period of its history, and that this annual gathering of representative men of science of Canada might well afford to stop in its haste to deal with specific achievements and local advances on some narrow sectors of the field, in order to consider the major operations, and to try to understand how the whole front line is swinging and what is the strength of the latest forces which are now operating on it.

We may discuss this subject from two quite different points of view; we may point out that this is a war waged by science, and may elaborate the applications of new and old scientific principles in the most intricate and devilish engines of offense and the equally ingenious and effective devices of defense, or in the most gruesome new modes of death and agony and the equally wonderful methods of preventing death, disease and permanent disablement, and of soothing and alleviating pain and misery. Much has been said on this both moral and unmoral status of science; fostered by us as the great servant of civilization and the promoter of prosperity and comfort to men, it is equally potent for the horrors of war, and the same power that has banished diseases and ameliorated the terror of wounds has poured out the deadly gas and high explosive on innocent women and children. To the devoted adherent of Science this is a most fascinating line of thought, and I am sure any one of us would be voluble upon the part his special science, or corner of science, has played in winning this war—for win we shall. The physicist would like to tell you of the rapid progress he has made in the mastery of a new element, the air, and the theory of ærodynamics; what ingenious acoustic devices he has invented to locate and destroy the lurking submarine and the hidden monster gun; to what lengths he has pushed the applications of wireless waves, of electrical instruments of detection; of the new uses to which he has put old apparatus. The chemist could recount the triumphs in the making of explosives, and the improvement of processes in their making, the production of deadly gases and gas protectors, the making of dyes and optical glasses, the cheapening or simplifying of methods of production, and the discovery of substitutes for many products.

But though this would be most instructive, and most interesting, even the parts of the story which are not secret and can be told to-day, yet I have preferred to take up the second point of view, namely, the effect the war has had, and is having, and can have on Science.

Perhaps the first result of this war on Science in British countries has been the confidence it has given us in our standing today and in our own accomplishments in the past. In general we can assert that the challenge to British science to show what it could do to meet the array of scientific devices elaborated in long secret by the enemy has been brilliantly taken up by our own scientific men, and I think we can truthfully say, we have gone them one better. We have proved to ourselves that British men of science possess a knowledge and a resource certainly not inferior to the German; and we have always known that we have had more than our share in the discovery of the great underlying principles and ideas and theories in the application of which Germany has been so persistent and so successful. The conviction that our chemists and engineers can achieve by our liberal methods results which we have been told could not compete with German drill-sergeant, dogmatic and cast-iron methods, has been of the greatest service to us. For I think it is true that we have felt instinctively that rather than fall to his level and follow in his footsteps, we would accept our losses and retain our individual freedom; indeed I doubt that we could ever have driven our scientific students and our people to this form of wooden subserviency to rigid and autocratic system. We have learned that co-operation, coupled with the retention of our freedom of effort and power of initiative, can accomplish all we need in order to wrest the supremacy once for all for our own people. It is a great thing for our scientists and our manufacturers and our capitalists to have achieved confidence in our scientific ability and strength; in two years England accomplished in armament and explosives in all their ramifications in every department of Chemistry, Physics, Optics, etc., what Germany took forty years to do.

Forced by the stress of war to a realization of the desperate condition in which she stood, on account of her past neglect of scientific method, as an organized state, England called at last to her aid her ablest leaders of science as well as her leaders of industry, appreciating late in life that the latter, great though they might be, were incapable alone of meeting and countering the peril that confronted the country. At last it was realized that research and its keenest prosecution alone could hope to match the results of German devotion to the fullest application of every known invention and experimental evidence which the most excellently equipped laboratories could produce. And, though pitifully few in numbers, and handicapped by the paucity of great industrial laboratories, the scientific men of Britain, as I have said, nobly answered that call, and their achievements became patent to the lowest and highest classes alike; and this is the greatest result that the war has had for us as scientists. The word *Research*, long known only to a small group of enthusiasts, has come before our people as a new watch-word; but it has taken a whole world in arms to write it in large enough letters so that our British eyes could see and read it. Before the war it had gradually been appearing to him out of the mists of prejudice and stupid conservatism; it had begun to reach large groups of the community to whom earlier it was entirely unknown. The farmer almost grudgingly admitted that those college fellows' advice was valuable; and the wonders of the results of research in modern medicine and surgery came to every household. It was when it began to reach his home and his pocket that science and research were raised in the estimation of the average man. Three years of war has accomplished for science what thirty years of peace might not have done. It is now the task and duty of the scientific bodies to take full advantage of their position, and to see that these words Science and Research carry their proper signification, and that they shall continue to receive after the war the attention paid them today. . . .

I do not wish here to lay too much stress on the difference between so-called pure and applied science; they are but two developmental lines of a common purpose. Though their methods are the same, their spirit and springs of action are distinct. We must cultivate both, for only then are both most obviously essential. It is the interaction of the ideal and the practical or utilitarian that spells progress; each urges the other on to greater success and achievement. Whether as purely academic men of science or as constructive engineers, we must preach from two texts; 1st, cherish research for its own sake in every laboratory that exists, whether in university or industrial establishment; 2nd, instal a works and scientific laboratory in connection with every industrial establishment—to perfect its products, to eliminate its waste, to utilize its by-products, to develop new products, to devise new processes, to find new uses for its products, in other words to apply science to the industry. New science will bring new applications; new difficulties will evoke new science.

As members of this Royal Society we are particularly interested in the development of research in Canada, both research in pure science and industrial research. The meagre facilities we possess for conducting research, both in man-power and in laboratories, is a disgrace to a country of the population and wealth of Canada. Compared with countries like Norway, Sweden, Denmark, Holland, Belgium and Portugal, which are of the same order of magnitude of population, and which we do not like to place above ourselves in intelligence, education and progressiveness, we find that Canada in research is a very poor second. We have not a single university properly developed for general research, and only a very few developed even in some departments. We can give all sorts of excuses for this state of affairs, but they are excuses, and the fact is not to our credit, as a people aspiring to full nationhood, that we must go abroad beyond our own borders to be educated. Any one of the countries I have named could have had about as good an excuse, if all their advanced students went to Germany. Within the last few years a few works laboratories have been established in connection with some of our largest manufacturing concerns, but very little research is, as yet, being carried on in them. Canada has nothing of the nature of the National Physical Laboratory of England or the Bureau of Standards of Washington; it has nothing national in education and science of any kind, except perhaps in agriculture; and yet the fostering of science is an absolute national necessity. The Research Council of Canada has made a visit to the United States to study this problem, and has seen the Bureau of Standards, the Bureau of Chemistry, the Bureau of Mines, the Carnegie Institution, the National Canners' Association Laboratories, the Mellon Institute, the Philadelphia Commercial Museum, and more of the same general type; and Canada has nothing of the kind whatsoever. Add to these the dozens of great universities' laboratories, a score of even larger research laboratories of manufacturing corporations, and hundreds of smaller ones, and one begins to have forced upon him the extent of the inadequateness of our preparation to meet post-war industrial competition, and of our present position as a parasite on the research institutions of our friends and neighbors. It is surely the part of this Section to urge upon the Government that it immediately vote the money needed for the foundation of some sort of National Research Institution for Canada. The opportunity is now, while the war has made our legislators see the value of Science and Research; lack of pressure now may leave us in the parasitic stage. What we are looking for is a better understanding and a feeling of mutual dependence between science and industrial enterprise, and a reliance upon ourselves and our own resources for our progress and our place in the industrial world. No state hereafter can be satisfied to be dependent on another for the essential products, when they can be produced within its own borders. These things mean that research must be stimulated, that the state must give the stimulation, that it must provide generously endowed research stations where these problems can be studied for the general good of the state.

The effect of the war upon Science should then result in industrial revolution. Its first effect on Canada should be seen in the stoppage of wastefulness and of the rapid destruction of our really limited natural resources, which are the very antithesis of anything connoted by the word scientific. This in itself would be an industrial and scientific revolution, and were it accomplished we should know the whole battle would be won. But it can not be accomplished without great improvement in fundamental scientific education. The problem of teaching science so that it is real, without at the same time falling into the other extreme, of devitalizing it and making it merely rule-of-thumb utilitarianism, is one that seriously confronts us. This is the second of those things which I think this Section should do, or take the lead in doing, to form a committee of themselves and others to give an intensive study to this problem for Canada, and formulate a national curriculum of scientific education for the schools and colleges of the land. It is through the rising generation that the hope for success lies, and I do not believe there is any more important national work that this national society can do than the making of a thorough survey of our special defects and needs in scientific education, and the urging of their conclusions on the educational authorities throughout the country, for it is my own experience and the opinion of many of my friends that science is badly and inadequately taught in Canada today. If the war will force us to remedy that, it will have accomplished another good end. It does not mean at all a new controversy or struggle with the Humanities; there is room for all, if all are well done. If every college fostered research and provided for it, and every student were thus brought into actual contact with the living and growing organism of science and felt its vivifying infection, the virus would be carried by him into the school or into the workshop or on to the farm or into the forest as a scientist, so that the idea of research would be ever in his mind as an agent of final reference in all operations and difficulties. Research would then come naturally into its own. We must rid the minds of the average being of the wonder and admiration and awe of science, which the popular magazine and newspaper cater to, and substitute familiarity and commonplace and solid understanding. We do not and shall not ask for science in place of the liberal arts, but science along with the other factors of the basis of knowledge which goes to the acquiring of the true art of living.

4 | Science, Education, and Research

The incorporation of science into the curriculum is but a part of the evolution of Canadian educational systems. One must indeed talk of different systems, for the British North America Act made the subsequent development of education primarily a provincial responsibility, and the results reflect correspondingly varied traditions and challenges.

Lower Canada inherited its educational philosophy and structure from the French régime, with only limited opportunities for elementary education. Convents and monasteries offered excellent education to the few, while the Seminary of Quebec provided higher education to future clerics in the classical tradition of the French Jesuits. After the English conquest, the level of schooling declined, and there was no satisfactory resolution of the debates about public universal education until the development of a provincial system of public schools after the union of the two Canadas in 1841. The roots of the problem were many and hopelessly tangled: diverging Anglican and Catholic efforts at controlling education, rivalry between Methodists and Anglicans, intransigent officials, the nature of Lower Canada's legislature, the Rebellion of 1837, social inequities, and, deepest of all divisions, cultural polarization between the English- and French-speaking communities. The bleakness of the situation is manifest in the Durham report of 1839.

By mid-century there were, however, prominent educationalists working for improvement, including the French-Canadian jurist Charles Mondelet and the future superintendent of public instruction, Jean-Baptiste Meilleur. Mondelet, Meilleur, and like-minded men were arguing for universal, public, non-denominational, and practical education, which would include a healthy amount of science and mathematics. Their ideal was not fully realized, but many of their proposals had been implemented by the time of Confederation. Although educational patterns varied from province to province, all shared the goal of universal primary education, and, especially for the English provinces, the United States furnished the model.

Higher education preceded widespread public education, and in Lower Canada was primarily the concern and creation of the Catholic clergy, who included science as an integral part of their classical curriculum. Thanks to their efforts, Lower Canada in the 1840s possessed the most extensive system of higher education in British North America, as was pointed out by Buller in an appendix to the Durham Report, and by George Young in his remarks on colonial science and education. French-speaking students in Lower Canada benefited from the co-operation of Church and government, while the English student was relatively poorly served, in spite of centres of excellence like the McGill medical school. Higher education in Upper Canada and the Maritimes was always the fruit of denominational activity. King's College in Toronto and the corresponding foundations in Fredericton and in Windsor, Nova Scotia, were projects of Anglican minorities. In both Upper Canada and New Brunswick these colleges aspired to the position of provincial universities, but long, vituperative debates arising from sectarian and social strife led to their both becoming non-denominational universities by the late 1850s. The debates of the 1850s showed increasing concern with practical education, and this trend was reflected in an expansion of the science curriculum and the eventual incorporation of engineering at the Universities of Toronto and New Brunswick. Victoria University, the Methodist college of Upper Canada, included science from its inception, thanks to its founder, Egerton Ryerson. McGill University, awakened from somnolence by William Dawson, principal from 1855, gradually developed a wide scientific curriculum. As we indicated in the first chapter, Scots or Scottish-educated professors feature prominently in the growth of science in English-speaking Canadian education: they were committed to science, eager to develop its practical aspects, willing to emigrate and also to accept lower salaries than their English counterparts.

A major emphasis in modern science education is the training of future scientists. This was not the case in the nineteenth century until the adoption of German patterns of university education. Science professors were not specialists in the modern sense, did not have laboratories, and rarely undertook research. Most colleges provided museums for natural curiosities, and, occasionally, a fairly complete range of geological, botanical, and zoological specimens. They also generally maintained cabinets of scientific apparatus for demonstration. Their aim was the diffusion of scientific ideas in a number of disciplines to as many students as possible, to diffuse a liberal education and an awareness of the wonders of God's creation. These were not the only objectives, however, for educationalists in Victorian Canada firmly believed that in the study of science, theory should be wedded to practice, and that this would contribute to the progress of all sectors of society. The view of science in Victorian Canada was fully utilitarian and full of heady optimism. Although each spokesman for scientific education for the truly liberally educated man had a slightly different view, our selection from the writings of Dawson and Ryerson illustrates a fundamental set of ideals that they held in common.

Confederation was followed by two decades of rapid educational advance and development. In many regions a student could then go from primary school to university and expect to get a reasonably good education that would normally introduce him to science and technology. Agriculture was put into school curricula by Ryerson in Ontario and by Ouimet in Quebec. Science and mathematics were to be found in high schools, government departments, and special technical schools. P.-J.-O. Chauveau was a major contributor to these achievements in Quebec. The curriculum of the Laval Normal School not only bespoke Chauveau's own interest in scientific and technical training, but reflected, after more than thirty years, the spirit of Charles Mondelet. A similar program could be found in the Ontario normal schools. Chauveau's admirable attempts were only partly successful. Despite his litany of important French-Canadian contributions to Canadian science, despite his pioneering attempts at developing technical and scientific education, French Canadians continued to excel in their traditional cultural spheres, while science and technology were predominantly within the English sphere. Chauveau himself was clearly aware of these cultural and professional differences.

Scientific research within the universities was largely a German creation of the early nineteenth century. The ideal of a professor as a researcher into nature rather than a passive inculcator of scientific ideas spread rapidly through German-speaking countries, and, with the aid of the Prince Consort, made some inroads into Britain in mid-century. In the United States the ideal of research and its university home, the graduate school, were late in capturing widespread interest. The same is true in Canada. Canadian universities did not move markedly towards research and higher degrees in science until the last two decades of the century. In a developing nation, the application of existing knowledge to practical ends was of more immediate concern than the extension of knowledge for possible future use. In times of prosperity and stability the case for research becomes stronger. At the turn of the century, Canada's scientific intelligentsia could look to German, French, English, and American models; the presidential address to the Royal Society of Canada in 1902 attempted to show the relation between university studies and scientific research, arguing that Canada would have to emulate the United States and Germany if she wished to move into the van of scientific nations.

The scientific and technical pressures generated by the First World War lent urgency to such pleas, and in Canada as elsewhere a national body was established to co-ordinate research. The Honorary Advisory Council for Scientific and Industrial Research was established in 1916 (see Section 3), and after the war was transformed into the National Research Council

(N.R.C.), designed not only as a co-ordinating body, but also for the allo-cation of Federal funds to support research. The creation of a National Research Institute for scientific and industrial enterprise was proposed in a way that led to considerable misunderstanding. Toronto and McGill seemed unduly favoured, and the Principal of Queen's University, R.B. Taylor, was outspoken in his criticisms. The difficulties can be seen in an N.R.C. memorandum, a letter from Principal Taylor, and a press release from the chairman of the N.R.C. The issue of government support for scien-tific research in the universities was subsequently resolved when the N.R.C. adopted the dual policy of maintaining its own research laboratories and providing funds for appropriate projects in university departments.

A. PATTERNS OF EDUCATION—LIBERAL AND PRACTICAL VIEWS OF SCIENCE

Lord Durham, *Report on the Affairs of British North America* (London, 1839), vol. I, pp. 133-6.

In the course of the preceding account, I have already incidentally given a good many of the most important details of the provision for education made in Lower Canada. I have described the general ignorance of the people, and the abortive attempt which was made, or rather which was professed to be made, for the purpose of establishing a general system of public instruction; I have described the singular abundance of a somewhat defective education which exists for the higher classes, and which is solely in the hands of the Catholic priesthood. It only remains that I should add, that though the adults who have come from the Old Country are generally more or less educated, the English are hardly better off than the French for the means of education for their children, and indeed possess scarcely any, except in the cities.

There exists at present no means of college education for Protestants in the Province; and the desire of obtaining general, and still more, pro-fessional instruction, yearly draws a great many young men into the United States.

I can indeed add little to the general information possessed by the Government respecting the great deficiency of instruction, and of the means of education in this Province. The commissioner whom I appointed to inquire into the state of education in the Province, endeavoured very properly to make inquiries so minute and ample, that the real state of things should be laid fully open; and with this view, he had with great labour prepared a series of questions, which he had transmitted to various persons in every parish. At the time when his labours were brought to a close, together with mine, he had received very few answers; but as it was desirable that the information which he had thus prepared the means of obtaining, should not be lost, a competent person has been engaged to receive and digest the returns. Complete information respecting the state of education, and of the result of past attempts to instruct the people, will thus, before long, be laid before the Government.

The inquiries of the commissioner were calculated to inspire but slender hopes of the immediate practicability of any attempt to establish a general and sound system of education for the Province. Not that the people themselves are indifferent or opposed to such a scheme. I was rejoiced to find that there existed among the French population a very general and deep sense of their own deficiencies in this respect, and a great desire to provide means for giving their children those advantages which had been denied to themselves. Among the English the same desire was equally felt; and I believe that the population of either origin would be willing to submit to local assessments for this purpose.

The inhabitants of the North American Continent, possessing an amount of material comfort, unknown to the peasantry of any other part of the world, are generally very sensible to the importance of education. And the noble provision which every one of the northern States of the Union has gloried in establishing for the education of its youth, has excited a general spirit of emulation amongst the neighbouring Provinces, and a desire, which will probably produce some active efforts, to improve their own educational institutions.

It is therefore much to be regretted, that there appear to exist obstacles to the establishment of such a general system of instruction as would supply the wants, and, I believe, meet the wishes of the entire population. The Catholic Clergy, to whose exertions the French and Irish population of Lower Canada are indebted for whatever means of education they have ever possessed, appear to be very unwilling that the State should in any way take the instruction of youth out of their hands. Nor do the clergy of some other denominations exhibit generally a less desire to give to education a sectarian character, which would be peculiarly mischievous in this Province, inasmuch as its inevitable effect would be to aggravate and perpetuate the existing distinctions of origin. But as the laity of every denomination appear to be opposed to these narrow views, I feel confident that the establishment of a strong popular government in this Province would very soon lead to the introduction of a liberal and general system of public education.

I am grieved to be obliged to remark, that the British Government has, since its possession of this Province, done, or even attempted, nothing for the promotion of general education. Indeed the only matter in which it has appeared in connexion with the subject, is one by no means creditable to it. For it has applied the Jesuits' estates, part of the property destined for purposes of education, to supply a species of fund for secret service; and for a number of years it has maintained an obstinate struggle with the Assembly in order to continue this misappropriation.

George Young, *On Colonial Literature, Science, and Education* (Halifax, 1842), pp. 180-8.

... Now that the Boundary question has been settled, and that the navigation of the River St. John has been secured to [the New England States] —that they are made competitors in the timber trade, as well as in the fisheries—that the West India ports are opened to the admission of their fish and lumber, at a moderate protecting duty—that our domestic manufactures have a protection against theirs to the extent only of 7 per cent— we may rely upon it, that, unless the education of the colonies is raised to a standard equal to that of New England, the lines of demarkation, so strongly, and almost offensively, marked by Lord Durham will be deepened, and the satires of Sam Slick be more clearly vindicated by the truth.

I open this Review with the following outline of the present state of Education in Canada, kindly furnished for the work by the Honble. Judge Charles Mondelet.

"A concise statement of the origin and progress of the leading Educational Institutions, will, it is thought, be conducive to the better understanding of the subject. Lower Canada is indebted for all its early scholastic endowments, to the liberality and zeal of the Catholic Church. The Jesuits' estates, the benevolence of that distinguished order, the same feeling which prompted the Seminaries of Quebec and Montreal, and of various Nunneries and their missions, laid the foundation, and gradually seconded the impulses which had been given. Had not the Jesuits' estates been diverted from their original destination, and their proceeds misapplied, the cause of education might have been much more advanced in this country than it has unfortunately been. ... From 1824 to 1836, various laws were passed by the Provincial Legislature, the enactments whereof, were modified yearly, until the Bill sent up by the Assembly in 1836, was lost in the Legislative Council. Each of those bodies asserted, as a matter of course, that they had sufficient grounds, the one for sending up, and the other for rejecting the Bill as framed. *Lower Canada has, until the last Session of the United Legislature, been left without any Legislative provision for popular education....*

"The following will suffice to give an *aperçu* of the history of the various educational institutions. The JESUIT COLLEGE was founded at Quebec, as early as the year 1635, through the exertions of a son of the Marquis de Gamache, who had joined that order. Until its suppression, the College was ably and successfully conducted by that distinguished Society. The URSULINE CONVENT, a Seminary for the instruction of young Ladies, in Quebec, was founded four years later, by Madame de la Peltire. . . .

"In 1663 the SÉMINAIRE DE QUÉBEC was founded by Bishop de Laval de Montmorency. When the Order of the Jesuits was suppressed by Clement XIV this college assumed the education of youth with complete success. Its present condition is very flourishing.*

"The means afforded to the English part of the community for classical and scientific education are smaller, that is, they have fewer institutions, although they are at liberty to send their children to the Catholic ones."

Upon this subject, Mr. Buller, in his Report to Lord Durham, hereinafter referred to, says:—

"With regard to the means of higher education, persons of British origin have hardly any, while those of French origin have them in too great abundance. It is impossible for an English gentleman to give his son a finished education in the province. If he wishes him to be instructed in the higher branches of mathematics, natural and moral philosophy, &c., he must either send him to Europe or the United States, or avail himself of the more imperfect opportunities afforded in the Catholic establishments of the colony. Political and religious animosities render them very adverse to the latter alternative. Some fear what they consider the contamination of republican principles

---

*Of this Institution Mr. Buller in his report . . . says 'the Seminary of Quebec is an admirably conducted establishment—the zeal of its members unremitting, and their arrangements in every way most judicious.' 'I had an opportunity last summer of examining this Institution under very favourable auspices. The buildings are extensive and commodious—the chambers airy and clean—food substantial and comfortable,— it has a valuable library—an host of Professors and Masters; and it secures to the student an extensive course of education. It combines both the Day School and College, for, in addition to the resident boarders, a large number of the boys of the City are taught here. I have before me the programmes for several past years. This course includes the English, French, Latin and Greek languages, Arithmetic, Grammar, Geography, Ancient and Modern History, Rhetoric, Logic, Moral and Natural Philosophy, Algebra, Mathematics, Astronomy, Chemistry, Music and the Art of Design. I spent some hours in the experimental Lecture Room of the eminent Professor Mon. Casault, and think that I saw there the best and most extensive set of philosophical apparatus, which is yet to be found in the Colonies of B.N. America. It is in daily use, and in a high state of preservation. The terms are moderate—the whole expence of a boarder not exceeding £25 to £30 per annum. Protestants and Catholics are admitted indiscriminately; and it is but due to the Catholic body who manage it, to state that no attempts are made at Conversion. I was promised a statement of the statistics and funds of this Institution before leaving Quebec, but it has not yet come . . . .'

in the States, and all shrink from the expense and separation attending education in Europe. Under these circumstances, they cherish with great fondness the hope of seeing the establishment of a Colonial University, on a broad and comprehensive scale. The better class of tradesmen, and the lower grade of merchants, are also without the opportunities of a good commercial education. It is true that there are some private establishments of the requisite description; but neither as regards number or quality are they adequate to the necessity."

"McGill's College, at Montreal, is now rising, and promises to become as distinguished as it is needed. Its beneficent founder, the late Hon. James McGill, has conferred on the community a boon, the benefits whereof will be daily more and more seen and felt."

The Report above quoted thus speaks of the state of this institution:—

"The only Protestant endowment in this province [in 1838] is that of McGill's College. The history of this institution, the original bequest, the protracted litigation, and, at length, the final decision, are matters as familiar to persons in this country acquainted with Canadian affairs as in Canada itself.—The college is not yet open; indeed, the building not yet erected. Its annual income, derivable from houses in Montreal, and money at interest, is about £644. It is obvious that this endowment alone, is insufficient for the purposes of a University, to which rank it is the wish of many to elevate this college; and it is doubtful whether the trustees of the Royal Institution, under whose direction it was placed by the will of the testator, would acquiesce in the terms on which legislative assistance ought hereafter to be granted."

It cannot therefore be said that this Institution is yet in effective operation. Means have been taken to engraft upon it a medical department of study,—conducted at present by seven medical men of the city of Montreal, Drs. Bouneau, Hall, Holmes, McCulloch, Campbell, Sewell, and Dick, who read lectures on the several branches of the profession. I was assured at Montreal last summer, that a superior course of medical education could be obtained here, and that some of the Professors were able and diligent teachers. The Special Council have aided this department by an annual grant of £500—and Sir Charles Bagot has promised at the present session of the Legislature, to recommend a permanent vote of £1000. In April last a project was published by some of the leading citizens of British origin and descent, for the establishment of an Academy to be called the "High School of Montreal." . . . In the Prospectus it is said:

"The great aim of the originators of the project for the establishment of a Seminary to be called 'The High School of Montreal', is to provide a system of Education for our youth, who are destined for the liberal professions or the higher walks of life and

business, upon a more comprehensive scale, and with greater efficiency in the practical conduct and administration, than can
possibly be attained in private Schools and Academies however
respectable. With this view they have been induced, after mature
and impartial consultation to give a decided preference to the general model of the best schools in Scotland, as being in their judgement, and without any disparagement to other Schools and
systems, best adapted both in their plan and working to the present condition of society in Canada.

"This will be readily admitted by all who are acquainted with
the characteristic merits of the Scottish system of Education.

"In the first place *it is eminently practical*, and fitted to qualify
those who go through its complete discipline and training for all
the offices and duties of active life. In the second place, *it is comprehensive and complete* in the range of the studies which it embraces. It gives no undue preference or disproportionate attention
to Classical, over Mathematical and Scientific learning. It gives to
each of the great branches of a liberal education its due place and
just proportion of time and culture. Another consideration that
had some weight in deciding this preference, is the greater facility
of obtaining eminent scholars, and able, faithful, and labourious
teachers, upon terms more economical from Scotland than from
any other of the sister Kingdoms."

*Report of the Commission Appointed under the Act of Assembly Relating
to King's College, Fredericton* (Fredericton, 1855), pp. 5-9.

TO HIS EXCELLENCY
THE HONORABLE JOHN HENRY THOMAS MANNERS SUTTON,
*Lieutenant Governor and Commander in Chief of the Province
of New Brunswick, &c, &c, &c.*

MAY IT PLEASE YOUR EXCELLENCY,
The undersigned Commissioners on King's College, Fredericton, have the
honor to report as follows:—

The Act of the Legislature, under the authority of which our proceedings
have been conducted, authorized the Governor in Council "to appoint a
Commission consisting of not more than five persons to inquire into the
present state of King's College, its management and utility, with the view
of improving the same, and rendering that institution more generally useful, and of suggesting the best mode of effecting that desirable object;

and should such Commission deem a suspension of the present Charter desirable, then to suggest the best mode of applying the Endowment, in the meantime, for the educational purposes of the Province."

Looking at the comprehensive terms of the Statute, and the Letter of instructions and suggestions addressed by His Excellency Sir Edmund Head to the Commissioners, the subject referred to the Commission appeared to embrace the whole system of Collegiate Education in New Brunswick; and accordingly, the undersigned Commissioners . . . proceeded during several days, to address themselves to the two-fold subject —as to what system of Collegiate Education is best adapted to supply the wants of the Province of New Brunswick—and as to whether King's College as now established is adapted to give effect to such a system.

First.—1. In considering the system of Collegiate Education best adapted to the circumstances of New Brunswick, we were unanimously of opinion that it ought to be at once comprehensive, special, and practical; that it ought to embrace those branches of learning which are usually taught in Colleges both in Great Britain and the United States—and special courses of instruction adapted to the agricultural, mechanical, manufacturing, and commercial pursuits and interests of New Brunswick; and that the subjects and modes of instruction in science and the modern languages, (including English, French, and German,) should have practical reference to those pursuits and interests.

2. New Brunswick would be retrograding, and would stand out in unenviable contrast with every other civilized country in both Europe and America, did she not continue to provide an institution in which her own youth could acquire a Collegiate Education such as would enable them to meet on equal terms, and hold intercourse with, the liberally educated men of other countries. New Brunswick would cease to be regarded with affection and pride by her offspring, should any of them be compelled to go abroad in order to acquire an University Education. The idea, therefore, of abolishing or suspending the Endowment of King's College, cannot be entertained by the Commissioners for a moment. On the contrary, we think there should be an advance rather than a retreat in this respect, and that the youth of New Brunswick, whether many or few, who aspire to the attainment of the best University Education, as preparatory to professional, or other active pursuits, should be able to secure that advantage in their native land.

3. The undersigned, therefore, recommend that a Collegiate course of instruction should be provided for, embracing the English Language and Literature—Greek and Roman Classics—Mathematics—Modern Languages—Natural History—Chemistry—Natural, Mental, and Moral Philosophy—and Civil Polity; that the standard of matriculation for entrance upon this course of study should be similar to that which has been established for matriculation in the University of Toronto; that the course of

study for the Bachelor of Arts Degree should extend over a period of three years; that the subjects of study and the system of options in pursuing them, for the appropriate exercise and cultivation of different useful talents, should be in harmony with what has been adopted by the most experienced and practical educationists in the recently established Colleges in England and Ireland, as well as in Canada.

4. But to provide for this class of Collegiate Students only, as has heretofore been the case in New Brunswick, and as has been the case in most Colleges in other countries, is to provide for only a small proportion of those youth who seek for the advantages of a superior education. The undersigned therefore recommend three additional courses of Collegiate Instruction...

5. The first of these special courses of study is that of *Civil Engineering and Land Surveying*—embracing English Language and Literature, Mathematics, General Physics, Chemistry, Surveying, Drawing and Mapping, Mechanics, Hydrostatics, Mineralogy and Geology, and Civil Engineering, including the principles of Architecture. In the study of the subjects of this course, there will be some option, according as the Student purposes to be a Land Surveyor or Civil Engineer.

6. The second special course of study is that of *Agriculture*—embracing the English Language and Literature, Chemistry, Elements of Natural Philosophy, Zoology and Botany, Theory of Agriculture, Physical Geography and History, Mineralogy and Geology, Surveying and Mapping, History and Diseases of Farm Animals, Practice of Agriculture, and Book-keeping.

7. The third special course of study is that of *Commerce and Navigation*—embracing the English Language and other Modern Languages, Arithmetic and Book-keeping, Physical Geography, Chemistry, Mathematics, Natural Philosophy, English Literature and History, Law of Nations and Commercial Law, and Navigation. In pursuing this course of study, the Student will be allowed some option in the subjects, according as he may intend to be a Merchant or Navigator....

10. By the courses of study thus sketched, and the facilities proposed to be afforded for attendance on single courses of lectures, the Commissioners are of opinion, that the higher educational wants and interests of New Brunswick are fully consulted; an University course of education comparable with that of any other country is maintained unimpaired for those who have the means and the noble ambition of acquiring general Collegiate Scholarship; while special and appropriate courses of instruction are provided for every young man who seeks to prepare himself thoroughly for entering upon any one of the great employments of agriculture—manufactures—commerce—land surveying—civil engineering—or navigation. Even any person who, with a view to some particular situation or branch of business, may feel it necessary to attend a single course of

lectures in Chemistry, Natural History, Natural Philosophy, Surveying, Engineering, &c. &c. &c., can avail himself of the advantages of College lectures for that particular purpose. Thus will the endowment and advantages of King's College be made available to every class of interests and of intelligent and enterprising young men in New Brunswick—to the Mechanic and Engineer, the Farmer and the Merchant, the Manufacturer and the Surveyor, not less than to those who seek the best preparation for any one of the learned professions.

11. In devising and maturing a proper system of University Education, the question of religious instruction has not failed to engage the most earnest attention of the Commissioners. On this subject there should be no difference of opinion in a christian land and among a christian people. No youth can be properly educated who is not instructed in religion as well as in science.

J.W. Dawson, *Science Education Abroad* (Montreal, 1870), pp. 12-15.

WANT OF SCIENCE TEACHING IN CANADA

Let us now turn to our own country, and study its means and appliances for the pursuit of practical science. The task is an easy one, for with the exception of two or three small and poorly supported agricultural schools, this Dominion does not possess a school of practical science. With mining resources second to those of no country in the world, we have not a school where a young Canadian can thoroughly learn mining or metallurgy; and as a consequence, our mines are undeveloped or go to waste under ruinous and unskilful experiments. With immense public works, and constant surveys of new territories, we have not a school fitted to train a competent civil engineer or surveyor. Attempting a great variety of manufactures, we have not schools wherein young men and young women can learn mechanical engineering, practical chemistry, or the art of design, or we are very feebly beginning such schools. We have scarcely begun to train scientific agriculturists or agricultural analysts. Our reasons for giving the necessary education to original scientific workers in any department, or of training teachers of science are very defective. Hitherto we have been obliged to limit ourselves to the provision of general academical courses of study, and of the schools necessary for training men in medicine, law and theology. Other avenues of higher professional life are, to a great extent, shut against our young men, while we are importing from abroad the second-rate men of other countries to do work which our own men, if trained here, could do better. Let us enquire then what we are doing in aid of science education, more especially in this commercial and manufacturing metropolis of Canada, which we may surely venture to regard as at least a Canadian Manchester, and something more important than a Canadian Zurich.

WHAT IS BEING DONE IN MONTREAL

(1) We have at least advanced so far as to regard physical science as a necessary part of a liberal education. In this University some part of natural or physical science is studied in each year of the College course, and we provide for honour studies in these subjects, which are at least sufficient to enable any one who has faithfully pursued them to enter on original research in some department of the natural productions and resources of the country, and to receive some considerable portion of the training which such studies can give. We have provided in our apparatus, museum, and observatory, the means of obtaining a practical acquaintance with several important departments of science. But in a general academical course of study too many other subjects require attention to allow science to take a leading place, and it is not the proper course of educational reform to endeavour to intrude science in the place of other subjects at least as necessary for general culture. We require to add to our general course of instruction special courses of practical science presided over by their proper professors, and attended by their own technical students.

(2) The lower departments of science education are to some small extent provided for by the teaching of elementary science in the schools. This, imperfect though it is, is of value, and I attribute to the partial awakening of the thirst for scientific knowledge by the small amount of science teaching in the ordinary schools in the United States and in this country, much of that quickness of apprehension and ready adaptation of new conditions, and inventive ingenuity which we find in the more educated portions of the common people. The Provincial Board of Arts and Manufactures also deserves credit for the attempts which it has made, under many discouragements, to provide science and art classes for the children of artisans. Proposals are also before the Local Legislature for Schools of Agriculture. The Local Government has procured reports on this subject from the Principals of the Normal Schools, and has also sent a special agent to study and report on the Agricultural Schools of France and Belgium, which are well worthy of imitation. A still more important suggestion has been made to the Dominion Government by the Director of the Geological Survey for the erection of a School of Mining.

These arrangements and proposals are valuable as far as they extend; but they fall short of providing the full measures of the higher science education, whether with reference to the training of original investigators, or of the various kinds of professional men required for the development of the resources of the country.

A. Egerton Ryerson, *Inaugural Address . . . Delivered at the Opening of Victoria College* (Toronto, 1842), pp. 15-17.

The knowledge of Mathematics being essential to the most lucrative pursuits, the study of that science has never been neglected or undervalued. The Arts of Navigation and Surveying, of Civil and Military Engineering, in all their various relations,—the two great national interests of Commerce and Internal Improvements, and the various departments of human industry,—are intimately identified with the knowledge of Mathematical principles, and are indebted for their present degree of perfection to the labours and researches of men profoundly skilled in Mathematical science. The influence of Mathematical studies, in disciplining and invigorating the mind, is not less important than the application of them to practical pursuits is advantageous and useful. The reasoning faculties are exercised and improved by the exactness of the science—its accurate, distinct, and infallible conclusions—and by the unlimited and certain discoveries of analysis; the former tending so essentially to strengthen the intellectual powers, the latter furnishing a most potent instrument for boundless research. Though the student, in after life, may seldom or never have occasion to refer to many of the Mathematical branches which he has studied, the habits of mind which they have contributed to form, will be advantageous to him in all subjects and pursuits which may engage his attention. Lord Bacon has remarked, that "men do not sufficiently understand the excellent use of the pure Mathematics, in that they do remedy and cure many defects in the wit and faculties intellectual. For, if the wit be dull, they sharpen it; if too wandering, they fix it; if too inherent in the sense, they abstract it: so that use which is collateral and intervenient, is no less worthy than that which was principal and intended." It was the opinion of Plato, that the youth thoroughly grounded in the Mathematics, would be quick and apt at all other sciences.

But the Physical Sciences—included in the *mixed* Mathematics—have, as yet, received little attention in our higher schools in this Province. Instruction has been chiefly confined to the Classics; and students have acquired little or no knowledge of Natural Philosophy, Chemistry, Mineralogy, Geology, Astronomy, &c., except what they have obtained in another Province, or in a foreign country. If one branch of education *must* be omitted, surely the knowledge of the laws of the universe, and of the works of God, is of more practical advantage, socially and morally, than a knowledge of Greek and Latin. How useful, how instructive, how delightful, to be made acquainted with the wonders and glories of the visible creation—the invariable laws by which the heavenly bodies are directed in their complicated and unceasing evolutions through the amplitudes of space—the structure of the earth on which we move, the materials of which it is composed, the arrangement of its component parts, the revolutions and changes to which its masses have been subjected, the laws which govern their ever-varying compositions and decompositions—the mechanical powers of air and water—the properties of light, heat, electricity, magnetism, &c.—and the application of these various branches of physical science to the arts of increasing the means of support, the comforts, refinements, and enjoyments of life, of facilitating the intercourse of nations, and of promoting the general happiness of our race! On these too-much neglected parts of a practical as well as liberal education, a vigilant attention should be bestowed, as physical science generally is nothing but the knowledge of nature applied to practical and useful purposes.

John Langton, 'The Importance of Scientific Studies to Practical Men', *Supplement to the Canadian Journal* (March 1854), 201-3.

Ladies and Gentlemen:—
I have been most anxious that the Library Association, which we have been forming in this town [Toronto], should also embrace a Mechanics' Institute; because, although a collection of books is an essential part of such an institution, a Library alone does not meet all the objects which I am desirous of promoting. A Public Library is designed to develop a taste for reading, and to afford facilities for the cultivation of literature generally, without a special preference for any particular department: a Mechanics' Institute, on the contrary, may in one sense be said to have a more confined object, being chiefly intended to promote the study of the Physical Sciences; but in other respects it embraces a larger field, by enabling its members to prosecute those studies, not from books alone but by the gradual accumulation of a museum of philosophical apparatus, and more especially by means of the delivery of Lectures. This being the end which I have been endeavouring to attain, I have been induced, with some other gentlemen holding similar views, to make a commencement with a short series of lectures; and keeping in mind that, which has been my leading object from the first, I have selected for the subject of the present discourse, the *Importance of Scientific Studies to Practical Men*.

It would be a waste of time, and almost an insult to your understandings, to enter into a formal defence of the uses and advantages of scientific knowledge. No such pleading can be required in the middle of the nineteenth century, when the last fifty years have witnessed a crowd of brilliant discoveries which have no parallel in history, except in the equally astonishing intellectual activity which distinguished the seventeenth century. But there are even now prejudices upon the subject, though of a very different kind from those which the first fathers of science had to combat, and which may deserve a word of comment. Within little more than fifty years from the dawn of modern science, the only true method of studying Nature was fully and firmly established and the foundation of most of the sciences was securely laid. The actual knowledge gained was mostly that of correct theory, and the opposition came from the learned, who had to forget their old doctrines and begin anew. The practical men hardly meddled at all in the disputes, or were on the side of the new discoveries. Now, on the other hand, the characteristic of the age is the practical application of our knowledge to purposes of immediate and obvious utility; and yet, curiously enough, it is from the practical men that the murmurs are chiefly heard.

One cause of this is, undoubtedly, the difficulty arising from the language of science, and the long and hard names which abound in scientific books. The very appearance of them repels the student, and he is apt to think that, were it not for the price of learning, they might as well be translated into his native tongue. The difficulty is, however, in a great measure unavoidable. Every trade and craft has its own peculiar technical terms, which are equally unintelligible to the bulk of society. A new fact, a new substance, a new system of classification must have its appropriate name. If you bestow upon it one already in use, and employed to designate something else, instead of rendering yourself more intelligible, you only create confusion. Every accession to our knowledge necessarily requires an addition to our vocabulary, and as science is for all nations, the new names are generally taken from those ancient languages which we have all equally inherited. *Carbon*, for instance, is taken from the Latin word for charcoal, and the chemist uses it as a name for that substance of which, with some trifling impurities, charcoal consists. If you translate it, and call it charcoal, it might seem more intelligible, but would really only lead you astray; for charcoal is only one of the forms in which we know carbon. It exists in almost equal purity in coke and in black lead (into the composition of which, by the way, not a particle of lead really enters), and in an absolutely pure state in the diamond. The element carbon is a new idea, and must have a new name. You cannot say, with truth, that a diamond consists of charcoal or of black lead, but all three consist of carbon. This new nomenclature may be, and is, perhaps, sometimes carried too far, and in such cases, everything that tends to give science an air of unnecessary profundity and obscurity should undoubtedly be amended. But, after all, the difficulty is not so formidable as it may appear, and at any rate it is a necessary evil; for you can no more speak of a science without using its language, than I could converse with a millwright about a saw-mill, without talking of "pitmen", "noddle-pins", "cross-heads", and "dogs", or with a sailor, without using such words as "shrouds", "dead-eyes", and "fids".

A more formidable prejudice is a sort of contempt which practical men sometimes entertain for theory. It is very common to hear a person spoken of as a theorist, in whom you cannot repose the same confidence as in a practical man; but we should not forget that a true theory is, as it were, only the essence of practice, or the generalization of a number of facts. And we cannot close our eyes to the numerous instances in which the greatest improvements in practice have originated in theory. Let us take an instance. There is, perhaps, no class more slow to yield an old prejudice than a sailor. Now it has been known, theoretically, for more than two centuries, that, to obtain the greatest advantage from the wind, a sail should exactly divide the angle between the direction of the wind and the ship's course; and it cannot do this unless the sail sits perfectly flat. If the sail forms a curve, only a part of it can be in the required position, and all the rest must be doing nearly as much harm as good. This was all known, but it was considered only theory, and sailmakers insisted that experience had shown that sails must be made to belly out, to catch, they said, and hold the wind. For two hundred years, practice would not listen to theory, till only the other day the prejudice was so far overcome, that the sails of a yacht were made as flat as canvas could be made to lie, and the consequence was that the America walked away from all her competitors. Old sailors will still no doubt shake their heads, but in the next generation a "bellying" sail will only be a poetical expression. Theory and practice, in truth, mutually assist each other. They are allies, not rivals; or you may liken them to a general and his soldiers. There were, doubtless, many men at Waterloo who could handle their bayonets, and go through their evolutions better than the Duke could have done; but they could no more have gained the battle of Waterloo without him to direct and combine, than he could have withstood a charge of the enemy without their collective strength.

One other objection to scientific studies I will mention, which always has been, and still is too common, especially among practical men. Where there is an obvious and direct application of some scientific truth or inquiry, the importance of the subject is willingly admitted; but where there appears no immediate prospect of turning it to account, the question is too often asked, What is the use of it? The objection is, in fact, more common now than formerly; for we have been so much accustomed of late years to witness the daily improvements in almost all arts and manufactures, that we are apt to undervalue everything that does not at once come up to our standard of utility. It cannot, however, be too thoroughly impressed upon a student, that no knowledge is without some use. As we say in common life, keep a thing for seven years, and you will find some purpose for it, so in science, a truth once ascertained is an accession to our knowledge, the importance of which can never be known till you can view it in connection with all around it. An anecdote is told of a celebrated sculptor whom a friend visited after a lapse of several days, and found still working at a statue that had appeared almost finished before. The friend wondered what he could have been doing, and the sculptor pointed out that he had scraped a little here and filed a little there, and brought out some feature more prominently in another place. "But these are only trifles," said his friend. "True," replied the sculptor; "but such trifles make the work perfect, and perfection is no trifle." So in science, a fact known is a stone prepared for the temple of knowledge; it may appear unimportant, and it may be idle for years, but time will assuredly show its proper place in the structure, and it *may* prove to be the keystone of an arch....

But it is said by some that you may leave such studies to the professionally learned, and that working men have no time for them: or if the nature of their occupation requires some knowledge of scientific results, it is sufficient for the mechanic to know the facts, and to work upon the rules which the philosophers have laid down for him. It has even been contended that the true principle of division of labour requires that the philosopher should devote himself to perfecting theory, and that the practical mechanic should confine his attention to attaining mere manual dexterity. To a certain extent this division must necessarily prevail, but if we are to look for much improvement in our present process, or much advance in our actual knowledge, the two branches must also be in a great measure combined. Theory and practice, as I have said before, mutually aid each other, and the mechanic cannot hope to attain much eminence without some theoretical knowledge, whilst the theorist must not disdain the aid of practical experience. The working mechanic, it is true, can but rarely become an accomplished philosopher, but he can at any rate become familiar with the principles of those sciences more immediately connected with his pursuits; and such is the mutual dependence of all the sciences, that he should at least have some idea of the general bearing and extent of our whole physical knowledge. A mere acquaintance with rules is not enough; for a man can never thoroughly understand, or even remember a rule, unless he knows something of the reason of it, and if he comes to apply it under slightly altered circumstances, he can never be certain that it continues to hold good for the case he has in hand. How many persons have wasted great mechanical ingenuity in attempting perpetual motion, which a slight acquaintance with first principles would have shown to be impossible! How many thousands have been thrown away in sinking shafts for coal, in strata which any geologist knew beforehand could contain none, or in working imaginary gold mines for what a mineralogist would, at a glance, have pronounced to be only mica! Again: if the object sought is possible, science will guide you in ascertaining whether the means used are sufficient for the purpose, or are the easiest, and most direct and economical, which can be employed. But more than all, theory will often suggest, and invite to a new track, which never would have occurred to a person unacquainted with science. In a word, if you are content to go on doing what preceding generations have done, you may perhaps trust to experience and rules alone; but if you wish to attempt anything new, where you can have no guidance from experience or rule, you must recur to first principles, which it is the province of science to teach. . . .

But if no practical mechanic can take full advantage of all the circumstances in which he is placed unless he have also some theoretical knowledge, neither can a mere theorist ever effect much who has not sufficient practical experience to know in what direction there is the greatest room for improvement, and what are the existing means for carrying it into effect. Almost all great discoveries and inventions have been made by men who united theoretical to practical knowledge.

J.W. Dawson, *On the Course of Collegiate Education, Adapted to the Circumstances of British North America* (Montreal, 1855), pp. 18-25.

We must now, however, direct our attention to the Physical Sciences, based on mathematical truth and on experiment; sciences which, independently of their intrinsic charms and value, have in our day established a connection so intimate with every department of mechanical, manufacturing and agricultural art, that without them the material welfare of nations cannot be sustained, much less advanced. I fear that the practical busy world scarcely yet recognizes this dependance of art on abstract science. Art, it is true, has often taken the lead of science and "developed results before their causes were understood;" but this is sometimes rather apparent than real, and on the other hand inventions which have their origin in scientific principles have become so rapidly diffused and so generally practised, that we are apt to forget the long series of investigations, the agitation of obscure scientific questions, and the indirect influences of even the doubts and difficulties of learned investigators, which have conspired to strike out the first hints of such practical applications. The more we enquire into this subject, the more will we be persuaded that the difference between the stationary condition of the arts in some ancient and modern semi-civilized nations, and their rapid progress among us, consists, to a great extent, in the more or less active pursuit and general diffusion of abstract science. Science has a double reward, first in the interest of its new facts and the ennobling general views to which it leads, and secondly, in its valuable and often unexpected applications. The long series of inquirers who from Galvani and Volta down to our time, questioned the occult and mysterious principle of galvanic electricity, were each rewarded by beautiful and striking discoveries, though they anticipated as little as the world that looked carelessly on their experiments, the result in that wonderful telegraphic communication, that now, in the hands almost of children, is at once the latest and greatest marvel of practical science, and a potential aid to commerce and civilizations. The scientific investigator and the academical professor may not be actual inventors; but they furnish the knowledge which leads to invention and they train the leading minds of society to appreciate and bring it into successful operation. Hence the school of abstract science is really one of the great moving powers in the material prosperity of nations.

Under this head it is unnecessary to refer to the importance of Mathematics as a means of rigid mental discipline, of industrial art, and of scientific progress; nor is it necessary even to name all those important branches of Physics which come under the denomination of Natural Philosophy. I rejoice to say that Prof. Howe, who has earned so high a reputation as the head of the High School, will in the present month, without, however, withdrawing himself from the oversight of the School, in which he is to have the aid of an additional master, assume the chair of Mathematics and Natural Philosophy in the College, and will as soon as possible commence a course of lectures on Physics, illustrated by the excellent apparatus of the Institution, which has been for some time lying idle.

Chemistry, whose claims are equally great with those of any department of Natural Philosophy, has not hitherto formed a part of the undergraduate course in this Institution, but it is hoped that, before next session, arrangements will be made to make the course now delivered in connection with the Medical Faculty, accessible to the students of arts in one of the sessions of their course.

I come now to the great group of sciences included under the name of Natural History, and comprising all that we can learn by the observation and arrangement of the works of creation, both in their present aspects and in those which they have presented in past time. Natural History, as cultivated in our time, is young and of rapid growth, and is even now only taking the place which its value as a means of training the observing powers and of enlarging our conceptions of nature, and as an auxiliary to industrial and fine art, demands for it. Zoology and Botany have for some time been necessary parts of medical education in many of the principal medical schools, and they will henceforth be accessible to students here. Geology and Mineralogy have been recognised by the governments of most civilized countries as important aids to material progress; and that they are so regarded here is witnessed by the admirable survey now in progress under my friend Mr. Logan, than whom no one would, I am sure, rejoice more in the diffusion of such a knowledge of his science as should render his labours more generally useful by making them better understood, and should increase the number of original enquirers. I hope before the close of the present month to commence a course of lectures on Natural History for the benefit of the students of the Medical Faculty and Faculty of Arts, and of such other persons as choose to avail themselves of it. I hope, also, in connection with this department, to form a Museum of Natural History, and shall be very thankful for any aid that may be given by individuals or public bodies toward such a collection.

Such is a very general view of the course of instruction adopted by us, and, as we believe, adapted to the present wants of this country, as a preparation for the learned professions and for general usefulness.

... We propose, then, to attempt the establishment of the following Special Courses, each to extend over two years, and to entitle the student, on examination, to a certificate or diploma.

1. A course of Civil Engineering. This will embrace English Literature, Mathematics, Natural Philosophy, Chemistry, Geology and Mineralogy, Surveying, and Civil Engineering, including the construction of machinery. Such a course will be exceedingly serviceable, not only to all young men about to enter on the profession of Civil Engineering, but to many others more or less closely connected with the public works or manufactures of the Province. In this department of Engineering we hope to enlist the talents of one of your Civil Engineers whose name is favorably known wherever the public works of Canada have been heard of.

2. We also hope to commence a course of Commercial Education, including English Literature, History and Physical Geography, Mathematics, Chemistry, Natural Philosophy, Natural History, Modern Languages, Commercial Law; and, if suitable arrangements can be made, Lectures on Political Economy. It is scarcely necessary to point out the advantages to the young men of Canada, and of this city in particular, which must result from the successful establishment of such a course.

3. A farther extension of our Courses of Study may be effected in the direction of Agriculture. Throughout the Colonies attention is now being directed to those scientific principles of farming which have effected such wonders in Great Britain, and the introduction of which is imperatively demanded in all the older and more worn out districts of this country. I have no doubt that there are within reach of Montreal a number of enquiring and intelligent young farmers, who would gladly avail themselves of such a course during the winter months. It would include the following subjects:—English Literature, Natural History, Natural Philosophy, Surveying, Agricultural Chemistry, Practical Agriculture, and Management of Farm Animals. ...

The present seems to be a time highly favorable for enterprise in the higher education of Canada. With natural resources and political institutions inferior to those of no part of the world, British America appears to have entered on a course of industrial and mental development whose results it is hardly possible to predict. The storms of party animosity which once convulsed these Colonies have to a great extent subsided into an honorable rivalry in the promotion of the great interests of the country. The highest public employments are open to the ambition of all; great public works and mining and manufacturing enterprises are calling for skilled labor; agriculture is passing from its first rude soil-exhausting stage to the rank of a scientific art; increasing population and wealth are constantly opening new fields for professional labor; the extension and improvement of elementary education are at once requiring higher attainments on the part of those who aspire to public positions, and offering to them the support of a more enlightened public opinion. The demand for educated men must thus constantly increase, and it is by fostering good collegiate institutions that this demand can be supplied in the best way— by training among ourselves the minds that are required.

G. Ouimet, *Rapport du Surintendant de l'Instruction Publique* (Québec, 1875-76), pp. xxvi-xxvii. (Translation.)

ON AGRICULTURAL EDUCATION

In our country, where most of the inhabitants are farmers, it is useful, it is necessary that the principles of agriculture be taught in all schools. There was a time when the Canadian soil, still new, yielded all crops without needing fertilizers, or special techniques of cultivation; but our land could not forever resist the debilitating regime to which it was submitted, and today people complain in many places that farming does not pay. It has thus become urgent to take steps to return the soil to its primitive fertility.

Ways of achieving this are known and within the grasp of everyone. Agriculture is an art that gave up its secrets long ago; we need only to popularize its principles. Schools offer the shortest path to this goal. Teach agriculture to the farmers' children, and agriculture will cease to be a blind routine.

In 1874, imbued with this idea, I prescribed this education for every school of the province. I had the good fortune to find an agricultural textbook that was both excellent and suitable for youthful minds; I am speaking of the *Petit Manuel d'Agriculture* by M. Hubert La Rue, which I provided for our primary schools.

But I am sorry to say that this tentative step was not crowned with the success I would have desired. About thirty thousand copies of the *Petit Manuel* were distributed, but our schools have more than two hundred thousand students. My measures were defeated by the apathy of a great many and by the ill will of some persons. Nevertheless, I hope that everyone will soon understand the importance of agricultural education, which is so closely connected with the dearest interests of our country, that it must become one of the first priorities in the annual grant. In any case, I shall formally order inspectors to require the teaching of the *Petit Manuel* to all students capable of understanding it.

Chauveau, *Instruction Publique au Canada* (Québec, 1876), pp. 132-4, 137. (Translation.)

The program of studies at the Laval Normal School is as follows:
Department of Male Teachers:

Third Year Students:
Religious instruction, logic, Latin grammar, Latin readings, Latin analysis, algebra, trigonometry, French dictation, literature, general history, theoretical and practical teaching, reading aloud.

Second Year Students:
Religious instruction, theoretical and practical teaching, French dictation, grammatical analysis, logical analysis, literature, mythology, geography, history of Canada, history of France, history of England, mental arithmetic, arithmetic, mensuration, algebra, geometry, astronomy, physics, chemistry, calligraphy, reading aloud.

First Year Students:
Religious instruction, theoretical and practical teaching, French dictation, grammatical analysis, Sacred History, history of Canada, mental arithmetic, arithmetic, weights and measures, geography, physics, agriculture, calligraphy, reading aloud.

All Students:
English reading, grammar, dictation, and grammatical analysis, translation from French into English, translation from English into French, solfeggio, piano and organ, military exercises, several lessons on natural history and on the uses of good fellowship.

Department of Female Teachers:

Second Year Students:
Religious instruction, history of the Church, theoretical and practical teaching, French dictation, grammatical analysis, logical analysis, litera-

ture, history of Canada, history of France, history of England, arithmetic, weights and measures, measuring, algebra, geography and the use of globes, agriculture, calligraphy, reading aloud.

First Year Students:
Religious instruction, Sacred History, theoretical and practical teaching, French dictation, grammatical analysis, literature, history of Canada, mental arithmetic, arithmetic, weights and measures, geography, calligraphy, reading aloud, agriculture.

All Students:
English grammar, analysis, dictation, reading, translation, piano and organ, solfeggio, drawing, hairdressing, knitting, flower-arranging, etc., etc.

The course of studies at the other two schools differs little from what you have just read.

Since their establishment, the normal schools have awarded 1,978 diplomas of which 128 were for academies, 759 for model schools, and 1,091 for elementary schools. These diplomas were obtained by 684 boys and 1,294 girls....

After this recapitulation, there are in all the institutions teaching above the elementary level, 948 students learning Greek, 1,656 learning Latin, 5,519 students whose native language is English learning French, 17,902 students whose native language is French studying English, 142 learning German, 742 intellectual and moral philosophy, 2,046 learning algebra, 2,226 geometry, 361 trigonometry, 346 differential and integral calculus, 685 physics, 455 chemistry, 1,727 natural history, 1,491 theoretical agriculture, 2,154 following the special commercial course, 2,199 learning linear drawing. We will not mention the multitude of other subjects that provide considerably smaller figures.

Chauveau, *Instruction Publique au Canada*, pp. 145-7. (Translation.)

[The Quebec] Parliament also voted a sum for the establishment of schools of science as applied to the arts, and decided to create three of them: one at Quebec, in connection with a Catholic institution, and two in Montreal, one of them connected with a Protestant, and one with a Catholic institution. McGill University had taken the lead and begun courses for which it subsequently received part of the grant. Laval University, at

Quebec, had also consented to offer this kind of education, and several of its professors had begun classes to that effect, but it abandoned the enterprise and returned the funds it had received to the ministry of Public Instruction. . . .

The reports of the two schools of science as applied to the arts testify to the rapid organization of these new institutions. That which is under the control of the commissioners of Catholic schools has taken the name of Ecole Polytechnique. It has four professors and fifteen students. The program of education is very complete and the system of examination, as indicated in the report, offers the best guarantees . . .

The costs of installation and the purchase of collections, libraries, laboratories, etc. have now risen to $7,405. The salaries of the professors for the year 1874-1875 were $2,160. Up to now, the entire expense ($10,710) has been completely defrayed by the government, with the exception of $1,710 paid by the commissioners of Catholic schools.

The school of the same type, opened under the auspices and direction of McGill University, has nine professors and 61 students. They have three distinct courses that cover three years each, and, in certain cases, two years of study, and they are adapted to the type of profession that the student wishes to enter. These courses are: (1) civil and mechanical engineering; (2) metallurgy and the exploitation of mines; (3) practical chemistry. The degree of bachelor of applied sciences, the degrees in civil engineering, and in applied science, are conferred upon students by the university. The total number of diplomas so far awarded is 36.

Two new schools of arts and trades have been opened under the direction of the Office of Arts and Manufactures in New Liverpool and Saint-Hyacinthe. The others are in Montreal, Quebec, Lévis, Sherbrooke, Trois-Rivières, and Sorel; giving a total of eight schools, 18 professors and 590 students.

The Montreal school has 293 students, of whom 134 have learned free-hand drawing, 29 architectural drawing, 48 mechanical design, 14 geometry, 20 modelling, 40 chemistry, 8 wash-drawing; a total of 234 lessons has been given by seven professors, and taking personal assistance into account, they have provided 5,516 individual lessons.

Charles Baillargé, *Technical Education of the People in Untechnical Language* (1894). (Read before Section III, Roy. Soc. Can., May 1894), pp. 1, 40-2. (Translation.)

While the Royal Society of Canada, 1894, like its prototype of England, cannot descend to the education of the masses of mankind; it behoves It though, as occasion may offer, to be suggestive of any improved mode of doing so. Persons not destined to become scientists, professors or even professional men or women, must nevertheless be educated to an intelligence of the phenomena of life; and it cannot be derogatory to the dignity of a society like this, to point out how it can be done without resorting to what is termed technology or technical tuition in the true sense of the word....

In the foregoing, the author assumes that men, otherwise uneducated, and, possibly, not even knowing how to read or write, either for want of means, or time or taste or aptitude for such tuition, are nevertheless curious or desirous of understanding the salient traits of the phenomena of life. He thinks they would be better for a certain amount of geometrical, physical, astronomical, geological, biological and other knowledge; more satisfied with themselves, less jealous of their educated fellows; less liable, when left to their own communings, to brooding in ill humour over their evils real or imaginary, of nourishing ideas of communism, anarchy and mischief; their minds being then stored with materials for reflection on the greatness and goodness of the Deity. The education of the masses should not go too far beyond mere reading, writing and arithmetic and the object-lessons here proposed. The injunction: "educate the people" must not be misinterpreted, as it certainly is in far too many cases. Education must be special to the calling of the individual, a truth which Lady Aberdeen is now trying hard to inculcate: servants, nurses, cooks must learn how to minister unto the wants of others. Too much grammar, literature, music, embroidery, conchology and the like, unfits the farmer's daughter for the duties of her husband's sphere in life. Collegiate, University tuition are ill adapted to the farmer's son, if he is to follow in the footsteps of his sire; though enough of chemistry to be imparted in a few object lessons and enough of botany are essential to his success in agri-, horti-, sylvi- and arboriculture. Let us beware of "too much education," there is a danger of overdoing the thing and thus causing or inciting our should be agriculturists to be dissatisfied with their parents' mode of livelihood, flocking towards cities and towns or populated centres; there to become second and third class professionals of every line, with little or nothing to do; with mischief and discontent and anarchical tendencies following in the wake. There must not be too much sacred history, too much catechism; Sunday is the day for that: it is so by divine injunction; let week days be devoted to object lessons in sewing, gardening, cooking, and in the teaching of husbandry: as how to dig, and ditch, and drain and plough and fence in, and cultivate, and erect buildings for the farm, the whole in miniature while at school, and in reality thereafter, for let it not be forgotten that God has said "six days shalt thou labour" and good, hard, honest work is the prayer the most congenial to the Deity.

B. SCIENCE IN THE UNIVERSITIES—THE MOVE TOWARDS RESEARCH

'The Universities in Relation to Research', *Proc. & Trans. Roy. Soc. Can.*, S2, 8 (1900), xlix-lix.

It is now many years since I came to the conclusion that the provision of adequate facilities for research is one of the prime necessities of university education in Canada; and it is with the object of accelerating the movement which has already begun in this direction that I have selected the relation of the universities to research as the topic of my remarks on this occasion.

It will, perhaps, be expedient for me at the outset to say that I propose to use the word research in its widest meaning, *i. e.*, as indicating those efforts of the human mind which result in the extension of knowledge, whether such efforts are exerted in the field of literature, of science or of art. It is a common mistake to apply the term research to what we somewhat erroneously denominate as "science," meaning thereby the physical and natural sciences. This limitation is comparatively modern, and science so defined is after all only a part of human knowledge.

The chief agencies of modern organized research are (1) the learned societies, and (2) the universities. The former receive and publish research papers; the latter superintend and direct the investigators and publish results. To these should properly be added the various journals which have been established and carried on by private effort. It is a significant fact that the establishment of modern learned societies coincides closely in time with the Renaissance movement. . . . Organized research in universities was of slower growth. In them the mediæval spirit was tenacious of life, and it was only in the nineteenth century, in Germany, at the close of the Napoleonic wars, that research, not only in natural philosophy, but in the whole field of knowledge, became the basis of the German educational system; and I might remark, without going into details, that the university systems of France and the other principal countries of Europe, with the exception of Great Britain, are in the main parallel with that of Germany, although not so consistently elaborated. To understand then what organized university research means in the fullest development which it has hitherto attained, let us turn our attention a little to Germany, of the educational system of which it forms an essential part.

We are so subject to the authority of words that it is difficult for us to realize that the organization called a university in Germany is almost entirely different in scope and object from the institution which we so designate in this country. Hitherto, at least in England and Canada, the function of the university has mainly been to impart a general and liberal education, continuing and completing the beginning already made in the secondary school. Speaking generally, I may say that under the German system the work of our secondary schools and universities combined is performed by the gymnasium, the nine or ten years' training of which leaves the young man of nineteen or twenty years of age with a much better liberal education than that possessed by the average graduate in arts of an English, Canadian or American university.

It is upon this substantial preliminary training that the work of the German university proper is based. Up to this point the young man has been a "learner," on entering the university he becomes a "student." This distinction, expressed by the German words "lernen" and "studieren," marks the difference between gymnasium and university—the acquisition of knowledge under the teacher in one, the independent research under the guidance of the professor in the other....

Let us now turn our attention for a few moments to the British university system. An extended description is unnecessary, since we are all familiar with the working of British universities themselves, or with the Canadian or American development of the original British type. Hence, it may suffice if I contrast briefly the British and German systems in some of their essential features.

In the organization of the German university research has been shown to be a fundamental principle; in the British university it is as yet incidental or of sporadic manifestation. I do not, of course, ignore the very important contributions which have been made by British scholars to the advancement of learning, but it is worthy of note that the credit for their splendid achievements is rather due to the individuals themselves than to the universities with which many of them were connected. The British university is not primarily an institution for research. In its function of providing the higher grades of a liberal education the proper comparison is with the upper classes of the German gymnasium, not with the German university proper. True, we find in some of the British universities a specialization in certain subjects, e. g., in honour classics and mathematics at Oxford and Cambridge leading to higher work than that attempted in the gymnasium; but however advanced the studies may be, there is rarely any attempt to guide the English undergraduate in the direction of research. Reading and examinations are the academic watchwords, and to the great mass of students and tutors the field of research is a terra incognita....

But the British nation is on the eve of an awakening, an awakening which has already taken place among certain leaders of thought. The fact is dawning upon the British mind that some vital connection really does exist between national progress and scientific discovery, and that the latter should be fostered in connection with the higher institutions of learning. Under the conviction that British commercial supremacy will be seriously threatened unless foreign, and especially German, scientific methods are adopted, universities of more modern type than Oxford and Cambridge, and also technical colleges, have been established. Such institutions no doubt fill a long-felt want, but they do not go to the root of the matter. On the academic side they are but a modification of the older type; on the technical side they contemplate, not the discovery of new truth, but the application of what is already known. The spirit of research is lacking, and without it no expenditure of money, no raising of examination standards for mere acquirement, will actually increase the capital account of national knowledge.

The policy of the universities of the United States regarding this matter is in marked contrast with the indecision and conservatism which prevail in the mother country. The type of mind which has been developed in the century and a quarter of separate national existence is one of great vigour and originality; but these qualities have for the most part been turned aside by the circumstances of a new country from abstract investigations. Research after the almighty dollar by the nearest short-cut has been, and perhaps still is, regarded as the chief national characteristic of our American cousins, and in this pursuit they have displayed a genius for concrete research in mechanical invention and an ability for commercial and industrial enterprise which have been an object of wonder, and latterly of anxiety to other nations. During the first hundred years of national existence the university of the gymnasium type which was inherited from England continued to develop and expand in the United States. Suddenly, however, almost exactly twenty-five years ago, a remarkable modification was introduced. The year 1877 marks an epoch in the establishment of the Johns Hopkins University, with research courses leading to the degree of Ph.D. as an addition to the usual undergraduate work; in other words, a grafting of the German university system upon the original stock. It is proper to state that even before that date research work had been prosecuted incidentally in some of the older existing universities. On consideration of the circumstances it is not difficult to account for this new departure. The movement was undoubtedly due to the influence of American students who had gone to Germany for special studies. This migration to and fro had been going on for some time before the founding of Johns Hopkins and still continues, the number of such students gradually increasing from 77 in 1860 to an average of about 400 annually during the last decade. The new university experiment was a success from the first. The scheme was carried out on such a high plane that large numbers of able and zealous students were attracted from all parts of the continent by the facilities for higher study and by the scholarships and fellowships which formed part of the scheme. The appointment of graduates of Johns

Hopkins to positions in other universities and their success as teachers and investigators have led to a widespread demand for professors who have proved their capacity for original work.

Since 1877 many other universities, including the best of those already in operation, as well as new foundations, have added a graduate department leading to the Ph.D. degree, although none of these, with the exception of Clark University, has made the prosecution of research the sole business of the university. Some idea of the rapid progress of this movement may be gathered from the fact that the numbers pursuing graduate studies in the universities of the United States have increased from eight, in 1850, to 399 in 1875, and to about 6,000 in 1902. We must conclude from these figures, I think, either that the national mind discerns some ultimate advantage in the cultivation of abstract science, or that, for once, it has been mysteriously diverted from the pursuit of the "main chance." It is surely significant that a practical philanthropist like Mr. Carnegie has recently bestowed the magnificent endowment of $10,000,000 for the establishment of an institution to be devoted solely to the promotion of research.

As to the ultimate scientific value of what has already been accomplished in the way of research under the influence of this recent movement, there is room for a qualifying remark. It must be remembered that much of the graduate work referred to does not mean actual research, the course for the Ph.D. in many cases being no higher than the honour B.A. course with us. What is required to remedy this unsatisfactory condition is that the Ph.D. be given only on the German plan, and that the main test therefor, a research, be published. When this condition becomes absolute there will be material for the world's judgment as to the amount and quality of the contribution to the advancement of knowledge.

Organized research in Canadian universities, as a definite system, can scarcely be said to exist as yet, although within the last decade certain beginnings have been made which indicate a movement in that direction. Canada, like the United States, has derived its university ideals from Great Britain. Some of the original faculties of our universities were a transplantation, so to speak, of groups of scholars from Britain, who brought with them intact the traditions in which they themselves had been nurtured, so that we received by direct importation scarcely more than fifty years ago a system which in the United States had been developing in its own way since the founding of Harvard in 1636. I cannot better illustrate the attitude towards research of many of these academic pioneers than by quoting the remark made by an English professor—himself a classical scholar—on an occasion so comparatively recent as the establishment of the physical laboratory in the University of Toronto. "Why go to the expense," said he, "of purchasing this elaborate equipment until the physicists have made an end of making discoveries?"

In the interval the idea of research has made gratifying progress among the well-informed. Probably few scholars could now be found in Canada who would put their objections so naively as my classical friend. This progress has come in part from a natural process of evolution within ourselves, and in part also from external influences, notably that of Germany and the United States. Many of our graduates have pursued courses of study in Germany and have brought back with them the German ideal. Besides, such is the geographical position of Canada with regard to the United States, and such the community of social and intellectual life, that the universities of these two countries must inevitably develop along parallel lines; and hence, if for no other reason, we may look forward to the gradual extension here of the research movement which is already so widespread in the neighbouring republic. . . .

We come finally to the effect of research upon the national life. Canada, it is true, is barely on the threshold of national existence, rich, however, in natural resources, and richer still in the physical, moral and intellectual qualities of its people. Its future as a nation will depend largely upon the aggregate of intellectual effort of its population. In this sense truly knowledge is power. The time has surely come when we should cease to take all our knowledge at second-hand from abroad, and when we should do some original thinking suitable to our own circumstances. Under the term original thinking I do not include merely the researches of the laboratory, for the spirit of research which inspires the chemist or the philologist is one with that creative faculty which moves the poet and the novelist, a spirit which guides all contemporary movements in literature, science and art. For the development of this spirit of originality the country must look primarily to its universities, for on them depends ultimately the whole intellectual life of the people. The time is approaching, if indeed it has not already arrived, when the research university must be regarded as the only university, and the task is incumbent upon those in authority of elaborating a university system not necessarily in imitation of those of other lands, but one which shall have proper regard to the importance of this new factor as well as to the past and future of our country.

'Memorandum Regarding the Communications of the Principal and the Registrar of Queen's University on the Proposed National Research Institute of Canada', in M.W. Thistle, *The Inner Ring* (Toronto, University of Toronto Press, 1966), pp. 51-8. Reprinted by permission of University of Toronto Press.

The communications on this subject addressed to the Acting Prime

Minister, Sir Thomas White, have been carefully considered by the Administrative Chairman of the Research Council, who desires to present a statement of the facts which bear on the question raised.

1. The University side of the Question was given prolonged consideration
The Council, brought into existence over two years ago, comprehends in its membership, amongst others the following:

Dr. A. Stanley Mackenzie, President of Dalhousie University, Halifax, N.S.

Dr. Walter C. Murray, President of the University of Saskatchewan

Dr. Frank D. Adams, Dean of the Faculty of Applied Science of McGill University

Professor R. F. Ruttan, Head of the Chemical Department of McGill University

Professor S. F. Kirkpatrick of the Department of Metallurgy of Queen's University

Professor J. C. McLennan, Head of the Department of Physics in the University of Toronto

These, with the Administrative Chairman, constituted seven of the total eleven members of the Council and this preponderance of representatives of the University point of view excluded . . . any neglect of the part which the Universities of Canada could and should play in the development of the Industries of the Dominion through scientific research and the application of the most advanced methods to industrial production. Some of these named have achieved distinction in scientific research and are, in consequence, known abroad as leading Canadian representatives of science.

2. The proper organization to promote industrial research in Canada
The question of the proper organization, or organizations, to promote scientific research in relation to the industries came before the Council at its second meeting, in January, 1917, and from May in that year till April of 1918, that is for a year, it was more or less constantly under consideration and discussion by the Council at its various meetings. Although the view that a National Research Institute should be founded developed before the end of 1917 and was supported by all members in attendance except one, the Chairman, regarding the question as of overwhelming importance and, therefore, to be exhaustively considered, prevented the Council from formally recording its decision on it until the meeting in April of 1918. From that decision which pronounced in favour of a National Research Institute there was only one member dissenting, namely, Professor Kirkpatrick of Queen's University, . . .

4. The reasons which influenced the Council
The reasons which prompted this decision are not far to seek, and one of

these thrusts itself upon consideration when the view advanced by Professor Kirkpatrick and now urged by Principal Taylor and Mr. Chown is examined. There are eighteen Universities, at least, in actual operation in Canada. All of these have teaching staffs in Science, more or less well supported, and all of these would desire to receive grants from the Dominion Government, were it decided to enter on a policy involving financial aid to the Universities. That such aid should be given only to a few which would observe strictly imposed conditions would work in theory, but in practice it would be quickly found impossible to exact these conditions and very shortly all would receive grants, corresponding in each case to the influence each could bring to bear on the political elements in the Dominion. The result would be inefficiency and an orgy of expenditure under no control or system, for an independent Board designed to exercise control would be subject to such pressure and criticism from interested quarters as to hamper seriously its usefulness and prevent it from doing its duty.

That this is not an imaginary danger may be gathered from the history of the university question on this continent and especially in Canada. It is invidious to dwell on this subject and, therefore, it will not be pursued further but what it signified could not be ignored by the Research Council. Is it surprising then that the members of that body who occupy university positions, with one exception only, are opposed to a policy which, if put in operation, would have the serious consequences referred to?

5. The proper functions of Universities in relation to science
The Universities of Canada should concern themselves with research in pure and applied science. These constitute the basis on which all industrial research must be laid. The superstructure must change because the foundations are always being altered, widened, extended or, now and then, in part, discarded. The ideal duty of the Universities is to labor on these foundations in order that they may be securely laid. That ideal is attainable and the Canadian Universities should concentrate all their energies to realize it. They have not done so expressly in the past. Sentiment, opinion and the forces that would make research in science one of the prime functions of a University have been lacking, and, in consequence, the contribution to new knowledge in Science they have made has fallen far short of what it should have been. A few members of the Scientific Staffs of several of the Universities have, in face of incredible difficulties and discouragements, achieved distinction in scientific research, but many more would have done so, had Boards of Governors, Presidents and those who have the power to formulate a policy on this line been not only sympathetically appreciative of research, but also energetically helpful in promoting it.

This is a situation that demands reform and it is of the first order of urgency. The Universities should be equipped and staffed to train the new army of researchers who are to assist in the application of science to Canadian industry. Only two of our universities offer, and inadequately, graduate courses involving research work in Science, and, as a consequence, there is today a great scarcity of properly qualified research men in Canada, a scarcity that is a serious handicap to efforts made to further the development of our industries on the scientific side.

Now, to thrust industrial research into our universities under these conditions is to aggravate the evils of the present situation. Industrial research is urgent, but it is fundamentally dependent on pure science and to provide for the former only on the supposition that the claims of pure and applied science are thus met is to make a profound mistake. It would be fatal to the best interests of our universities. This is true also for the situation abroad and in support of this one could cite the views of many leading scientific men who, because of their experience, are entitled to be heard on this subject, but it will suffice to quote from the address delivered in November last by Sir J. J. Thomson, as President of the Royal Society of London:

> To increase the resources and equipment of the Universities would, I think, be the most effective way of aiding research in pure science. If the grants for this have to come from a fund which has also to provide these for industrial research, there is, I think, no inconsiderable danger that the latter may be regarded as the more urgent and that the claims of pure science may be crowded out.

Is it any wonder then that a broad state aid for industrial and technical research is provided for almost wholly in institutions apart from the universities? Germany, whose rulers were in this matter, as everyone will admit, eminently practical, fostered research in pure science in her twenty-three universities and industrial and technical research in several institutions absolutely independent of the Universities. In Great Britain the National Physical Laboratory, which concerns itself with problems of technical research, and has an annual budget of more than $750,000, is not connected with any of the British Universities. The Bureau of Standards at Washington, the National Laboratory for technical research, is not affiliated with any University in the United States. The foundation of a National Research Institute in France is under consideration by the French Government. The Australian Advisory Council of Science and Industry has recommended the foundation of a Research Institute for Science and Industry, which is to be independent of the five Universities of the Commonwealth. In Japan a fund of $2,600,000, $1,500,000 of which has been granted by the Emperor and the Diet, has already been gathered for the establishment of a National Research Institute which is to be unconnected with any of the four Japanese Universities. What is there in the situation in Canada that would justify a disregard of all this experience and trend of opinion?

6. The best way in general to aid industrial research
The most defensible policy in the relation of the State to industrial research is that which would put the greater part of the burden of the expense of industrial research on the industries themselves. Unfortunately, all the firms in each line of industry in Canada cannot individually and independently carry on research, but by combining their resources as Guilds for research they could provide effectively for it if it were conducted under the supervision of a highly trained staff such as would direct the work of a National Research Institute as proposed. The only expense that the State would be put to would be that for the accommodation, light and heat in the laboratories of the Institute. This would bring together under the same roof industries with common or related problems, which would insure coordination and cooperation in the researches undertaken to solve them. The results would be of service to all firms in the lines of industry concerned. The country as a whole would benefit thereby. Without this coordination and cooperation there would be over-lapping of effort and greatly increased expenditure and the result would fall far short of what it should be. To scatter these guilds in the Universities would put a premium on lack of cooperation and unwise expenditure and promote unideal relations amongst the Universities themselves.

When the Institute proposed is founded, individual firms will with their own resources, as now, be free to put their problems for research where they choose, but, as they will wish to keep secret or to patent the results, if they are of value industrially, why should the Dominion Treasury be expected to pay a cent of the expense incurred? From the public point of view there is no objection to any of the Universities undertaking this work for individual industrial firms for adequate financial rewards, although it may be urged that when, as stipulated in such arrangements, the results shall be kept secret even from students, or patented, this is not fulfilling the ideal function of a university.

7. Objections to Government Control of Industrial Research
It is urged against the proposed National Research Institute that, as it will be a Government organization, it will come under bureaucratic control, and, like the Scientific Departments of the Government, give unsatisfactory results. The Scientific Branches of the Dominion Government have not done all that they should but they have done excellent service for the

country as a careful survey of their contributions to Science shows. It is doubtful if politics ever played any part in the control of these organizations and the sins of omission and commission, of which they have been guilty, have been due to other defects. In any case, it is not quite appropriate for representatives of the Universities to indulge in criticism on this score, seeing how remiss in this matter they themselves have been in the past. Who, if he were bold enough and willing to be specific in his statements, could not point out many teachers in the Universities who so far as scientific research is concerned, to use Mr. Chown's phraseology, "petered out" as soon as they received permanent university appointments?

There should be no difficulty in creating the right organization to control the proposed National Research Laboratory in order to make it thoroughly efficient in its service to industrial research, an organization which will keep it free from politics, red tape and which will keep in close touch with the industries of Canada to meet their needs and to do so with no more expense to Dominion Treasury than is absolutely necessary. The National Physical Laboratory of Great Britain, which is under the control of the Council of the Royal Society, does splendid work in the way of research. The Bureau of Standards at Washington, which is under an Advisory Board associated with the Department of the Interior, has a record of service in Scientific research which speaks for itself. The creation of a proper governing Board for the proposed National Research Institute is not, therefore, an impossible task.

8. The Universities and the proposed Research Institute

The proposed Research Institute will concern itself with all measures to promote industrial research and it will have, besides, the function of investigating and thereby determining all technical standards used in the industries and the technical trades in Canada. It will endeavour to investigate the raw materials for new as well as already established industries and it will strive to lead the way in bringing the most advanced scientific methods to bear on the utilization of the natural resources of the Dominion. The Institute will not concern itself with the research problems of single firms in any industry which can and ought to undertake the total expenses involved. These will, if they think fit, resort to other organizations, Universities included, to have their problems solved or to have new methods and processes provided in their industries, or they may, as some of them do now, maintain their own research laboratories. The existence and operation of the Research Institute will not in the least affect these or their freedom to engage . . . in contracts with University Organizations to do their research work for them, as in the case of the

M. J. O'Brien Co. Ltd. and Queen's University. The Universities will, therefore, be as free in this matter as ever they were.

The true functions of the universities as regards research are too well known to require elaboration here. They should concern themselves not only with preparing young men in the lines of science before they enter on a course of training in research, but they should equip themselves to give effectively this training not in industrial research but in pure or applied science which is held by all who have experience in research to be the best field in which to prepare students for a career in industrial research. The problems in pure and applied science are clean cut, of general and permanent interest and their solution may have far reaching industrial application while those which concern industrial research though they are of importance and, therefore, to be undertaken, are, in the great majority of cases, narrower in scope and are not, therefore, likely to develop fully the capacity for research in the student. Of the vast majority of the problems of industrial research some are of local interest only, others there are whose solution may lose value because of some discovery in pure science, while others of them are insoluble until discoveries in pure science make the solution available. On the general principle of utility as well as because of ideals the student who is training for industrial research should during that period concern himself with problems in pure and applied science.

It is the duty of the universities to train such students and to train them well. To train them inadequately, thereby giving them a false ideal and a very limited outlook, is to defeat the hopes of those in the industries who are looking to scientific research to help them in the new and trying conditions which confront those industries. When an individual firm provides a laboratory for research, those engaged to staff that laboratory should be genuine, highly trained researchers and not amateurs or tyros in research.

To produce these highly trained researchers is the urgent duty of the universities. If the Canadian Universities will not produce them, then they must be brought from the Universities of the United States.

The Administrative Chairman regards this duty as one of urgency and he is led to emphasize it because of his knowledge of the present conditions in Canada regarding industrial research. His experience, derived from research as a biochemist of a third of a century, and from crusading, in season and out of season, for research in Canada all that time entitles him to an opinion on the subject opposed to that of very estimable and well-meaning gentlemen who never carried on research and whose interest in the subject dates, as it were, from yesterday.

9. The Question of the Hour

In concluding this statement, the Administrative Chairman would emphasize, on behalf of the Research Council the urgency of founding the proposed National Research Institute and that the interests of the Dominion as a whole should submerge any sectional interests.

R. Bruce Taylor to R.F. Ruttan, 27 Feb. 1919, in M.W. Thistle, *The Inner Ring*, pp. 62-7.

My statement regarding the Advisory Council that, "While Toronto and McGill had been given every consideration in the scheme, Queen's had been slighted", is unfortunately an exact statement of the fact and nothing is to be gained by any reticence in the matter. An effort is being made to launch a large scheme of research, but in doing so one of the three great Canadian Universities is being treated as though it had no existence. In proof of this statement I would draw your attention to the following facts:

1. When the Advisory Council was instituted there were included on it, together with representatives of McGill and Toronto, President Murray of the University of Saskatchewan, and President Mackenzie of Dalhousie, Queen's was entirely omitted. It was only when strong representations had been made at Ottawa, through Mr. W. F. Nickle, M.P., that Prof. Kirkpatrick was included as a representative of Queen's. If the Council wished to carry the support of Queen's along with it, this beginning was certainly unfortunate. That this omission of Queen's was not a mistake, but a matter of deliberate policy, is clear enough from what has followed.

2. In a letter to Prof. A. L. Clark, of date 9th Sept., 1918, Dr. Macallum says: "The Government will not give money directly, or through the Research Council, to the Universities to assist them in developing research. It holds that this is the duty of the Provinces which control education in their own spheres. There is no doubt whatever about the attitude of the Government on this subject". However, in a signed statement, contributed by Dr. A. B. Macallum to the Globe of 2nd January, 1919, the following indication of the trend of affairs was given. "A further necessary step will be the working out of the Council's plans for more adequate provision by the Universities for the training of qualified scientific workers. In the more generous investment of State funds for this purpose, starting say with Toronto, McGill, and L'Ecole Polytechnique, lies the hope of securing for the ensuing years of the world's strenuous and pitiless trade warfare the nation's leaders in scientific and industrial research". Why should L'Ecole Polytechnique have been grouped with Toronto and McGill, and why should Queen's have been omitted? Queen's has contributed to Metallurgical research in Canada certainly as much as McGill and a great deal more than Toronto. Research has always been carried on in Queen's

and so far as I know it has never been carried on in L'Ecole Polytechnique, why then was the French School included? Because if anything was to be done for McGill something had to be done for L'Ecole Polytechnique. It was not good policy to exclude the French Canadian element. I admit at once that the political difficulty is great, but the trail of politics has been over this scheme from the outset with the result that an injustice has been done to Queen's.

3. Again, in a statement made by Prof. Macallum at the Royal Canadian Institute on the 22nd. of November last, he is reported to have said: "One of the handicaps of this country is that there are few scientific researchers. If the work must be developed in Canada scientific men will have to be trained. Up to the present there has not been any particular training institution for these men, but Dr. Macallum intimated that part of the programme of the Advisory Council includes the equipment of at least two Universities for such work". The implication of that statement again is that Queen's is being passed by.

4. Prof. A. L. Clark, Head of the Department of Physics in Queen's wrote to Dr. Macallum on 1st. September last regarding some work on the resistance of carbon as used in telephone transmitters, and also on the subject of electrical double layers. No acknowledgement of the letter has been made.

I am well aware that the present condition of research in the Canadian Universities is unsatisfactory. Men are underpaid and overdriven, and the merely instructional side has undue importance given it. But this is a condition that can be remedied. The Advisory Council by drawing the attention of the Governing Boards of Universities to the matter has undoubtedly done something to awaken the Universities, and it can do a great deal more by encouraging the men within their walls. If men are to be trained in research, research must not be divorced from the Universities, and transferred to a Government bureau in Ottawa. Such a move would make the position worse than it is at present. The Bureau would inevitably, things being as they are, become part of the political machine, and appointments to it would become a matter of influence. The quiet worker would hardly have a chance as against the man who knew the ropes and condescended to pull them. There can be no analogy between such a bureau and the Mellon Institute, free from pressure from without.

However, these matters are beyond the scope of your letter. I think I have made out my case that Queen's has been slighted. I said nothing about money grants from the Advisory Council; I feel sure that no resolutions discriminating against Queen's have been passed by the Council. But the superior person has been abroad in the land, and Queen's resents the implication that in the matter of research it is not worth considering.

Statement to the Press by A.B. Macallum, March 1919, in M.W. Thistle, *The Inner Ring*, pp. 65-7.

PRINCIPAL BRUCE TAYLOR AND THE "GRIEVANCE" OF QUEEN'S UNIVERSITY

Absence from Ottawa for nearly a week and pressure of duties have prevented me from replying earlier to Principal Taylor's letter of March 3rd. on the "Grievance" that Queen's feels through alleged slights offered it by the Research Council or myself. . . . Dr. Taylor is solicitous about the danger of politics invading the Council or its organizations. In the more than two years of the Research Council's existence, and since the attempt of the functionary referred to there has not been the slightest effort made by either a Minister of the Crown, a member of Parliament, or a Senator, to affect the work or influence the decisions of the Council. Perhaps they knew they could not do it, for every member of the Council would resist to the utmost the introduction of politics, but I modestly fancy that they had not the slightest desire on that score.

Even if all the alleged "facts" which the Principal gives in his letter were correct, they would weigh as nothing in the issue which crowds into the situation. This issue is: Shall Canada have a National Research Institute like the Bureau of Standards at Washington, to aid in the utilization of the natural resources of Canada and to foster the application of science to our industries? Or, Shall the money which might be available for such an Institute be divided amongst the Universities of Canada to enable them to do this work? That is the whole matter in a nutshell.

Eight out of nine members of the Council, after the most protracted consideration of all the facts bearing on the question, decided to recommend the establishment of a National Research Institute, and six of the eight hail from five Universities, while the one opposed to this decision was a Professor in Queen's. As these are all scientific men, their decision should count for more than the opinion of Dr. Taylor who has had no training or experience in Science. Prof. Kirkpatrick at first maintained that all the money available for research should be divided amongst the Universities on some system of control, but he subsequently admitted that, in addition, a Bureau of Standards should be established. . . .

There are eighteen Universities in Canada . . . about the same number as Great Britain and Ireland have with more than six times the population of Canada. All the Canadian Universities are in a more or less necessitous condition, and, if grants from the Dominion Treasury were going, they would all scramble to get them. No system of check or control could be imposed for it would be swept away under pressure from the forces behind the Universities. Each of them not in receipt of money from the Dominion Treasury, or receiving, as it might think, not enough, would feel "slighted" and there would result a resort to all the arts that characterize the most undesirable kind of politics. Charges of unfairness and of the exercise of favoritism would be bandied about amongst the Universities just as even now Principal Taylor charges unfairness in having Dalhousie and Saskatchewan represented in the Research Council as originally constituted, while Queen's was not. There would be an expenditure on the Universities along this line that would far exceed what would be required for the maintenance of a Research Institute, and there would be very little to show for it in the way of increased industrial efficiency.

All the while research in pure Science, which is a primary function of the Universities, would be neglected or abandoned. The training of the researchers who are needed so much in the industries, and of whom there are so few in Canada, would not be encouraged, for such training is given best in pure Science, which, as Sir J. J. Thomson recently pointed out, would be in danger of being crowded out of the Universities by industrial research. This, of course, Dr. Taylor does not recognize, but the scientific members of the Research Council recognize the danger ahead, and hence their position on this question.

The Universities of Canada must, in the opinion of the Research Council, look to other sources than the Dominion Government for the funds which they require to provide for advanced scientific teaching and for scientific research. Where these sources are located must be left for each University to determine. They can and should make arrangements whereby the scientific members of their staffs will have much more time for research than they have now. This alone would enormously enhance the research output of these Universities and it would stimulate the enthusiasm of a large number of students who would prepare to enter on a career of scientific research. For such students the Council has established Studentships and Fellowships, which will enable them to achieve their object. It gives grants to University men and researchers generally, who have problems which can only be investigated on the provision for such financial assistance. It is organizing to assist the industries on the scientific side to enable them to hold their own in the markets at home and abroad. . . .

. . . The Dominion faces the task of providing an annual revenue of nearly four hundred millions of dollars and this revenue can be raised only if the country is very prosperous industrially. Scientific Research and the application of the most advanced scientific methods to the industries are therefore of paramount importance if the country is to succeed in raising this revenue. If it does not succeed, what is ahead? A National Research Institute, therefore, to assist the industries in adapting themselves to the situation now confronting them and thereby to help them to bear this enormous burden of taxation, is now of the first order of urgency. This is the hour for vision and all the true friends of Queen's earnestly desire that she will align herself with all those who share it. That is the earnest wish also of the writer.

BOOKS FOR FURTHER READING

The following list is not intended as a bibliography, nor as an indication of the major works on science in Canadian history. There is not one satisfactory general work on the subject. The works below will, however, provide starting points for those who wish to read more about particular issues raised in this volume. Many of the books listed here contain specialist bibliographies, and will be correspondingly more useful to the reader. The journals and books from which our readings were selected will of course repay consultation, and to facilitate this we have given precise references in the body of the text.

A. GENERAL WORKS (General Background)

Bourinot, J.G. *The Intellectual Development of the Canadian People* (Toronto, 1881).

Brebner, J.B. *The Explorers of North America, 1492-1806* (London, 1933).

Butterfield, Herbert. *The Origins of Modern Science* (London, 1950).

Cambridge History of the British Empire, vol. vi, *Canada and Newfoundland* (Cambridge, 1930).

Careless, J.M.S. *Canada, A Story of Challenge* (Cambridge, 1953).

Clark, S.D. *The Developing Canadian Community*, 2nd ed. (Toronto, 1968).

Creighton, Donald. *Dominion of the North* (Toronto, 1957).

Gillispie, Charles C. *The Edge of Objectivity* (Princeton, 1960).

Glazebrook, G.P. de T. *Life in Ontario. A Social History* (Toronto, 1968).

Hall, A. Rupert. *The Scientific Revolution* (London, 1954).

Innis, M.Q. *An Economic History of Canada*, 2nd ed. (Toronto, 1943).

Kearney, Hugh. *Science and Change 1500-1700* (New York/Toronto, 1971).

Reingold, Nathan. *Science in Nineteenth-Century America. A Documentary History* (London/Melbourne/Toronto, 1966).

Smith, Preserved. *A History of Modern Culture*, 2 vols.: vol. 1 *Origins of Modern Culture 1543-1687*; vol. 2 *The Enlightenment 1687-1776*, 1st ed., 1934 (New York, 1962).

Stanley, G.F.G., ed. *Pioneers of Canadian Science* (Toronto, 1966).

Tory, H.M., ed. *A History of Science in Canada* (Toronto, 1939).

Wade, Mason. *The French Canadians 1760-1967* (Toronto, 1968).

Warrington, C.J.S., and Nicholls, R.V.V. *A History of Chemistry in Canada* (New York, 1939).

Westfall, Richard S. *The Construction of Modern Science: Mechanisms and Mechanics* (New York, 1971).

B. NEW FRANCE AND THE VOYAGES OF EXPLORATION (Section 1)

Beals, C.S., and Shenstone, D.A., eds. *Science, History, and Hudson Bay* (Ottawa, 1968).

Dodge, E.S. *The Polar Rosses* (London, 1973).

Gosselin, Auguste. 'Les Jésuites au Canada: le Père de Bonnécamps, dernier professeur d'Hydrographie au Collège de Québec (1741-1759)', *Mém. S.R.C.*, II ser., sect. I (1895), 25-61.

Huard, V.-A. *La vie et l'oeuvre de l'abbé Provancher* (Québec, 1926).

Hughes, Thomas. *History of the Society of Jesus in North America* (London/New York, 1907).

Lamontagne, Roland. 'Les échanges scientifiques entre Roland-Michel Barrin de la Galissonière et les chercheurs contemporains', *Rev. Hist. Amér. Franç.*, 14 (1960), 25-33.

Ouellet, Cyrias. *La vie des sciences au Canada français* (Québec, 1964).

Rousseau, Jacques. *Jacques Cartier et la Grosse Maladie* (Montréal, 1953).

Roy, Antoine. *Les Lettres, les sciences, et les arts au Canada sous le régime français* (Paris, 1930).

Vallée, A. *Michel Sarrazin* (Québec, 1927).

Vallée, A. 'Cinq lettres inédites de Jean François Gaultier à M. de Réaumur de l'Académie des Sciences', *Proc. & Trans. Roy. Soc. Canada*, S3, 24 (1930), 31-43.

C. SCIENCE AND GOVERNMENT, SCIENTIFIC SURVEYS (Section 2)

Adams, Frank Dawson. 'Biographical Memoir of Thomas Sterry Hunt 1826-1892', *Biogr. Mem. Nat. Acad. Sci. 15* (1934), 207-38.

Alcock, F.J. *A Century in the History of the Geological Survey of Canada* (Ottawa, 1947).

Clark, T.H. 'Sir John William Dawson, 1820-1899' in G.F. Stanley, ed., *Pioneers of Canadian Science* (Toronto, 1966), 100-13.

Collard, E.A. 'Lyell and Dawson: A Centenary', *Dalh. Rev. 22* (1942), 133-44.

Geikie, Archibald. *The Founders of Geology* (New York, 1962; reprint of 2nd ed., 1905).

Harrison, J.M., and Hall, E. 'William Edmond Logan', *Proc. Geol. Ass. Canada 15* (1963), 33-42.

Kennedy, J.E. 'The Early Days of the First Astronomical Observatory in Canada', *Journal of the Royal Astronomical Society 49* (1955), 181.

Lefroy, John Henry. *In search of the magnetic north: soldier-surveyor's letters from the North-West, 1843-1844*, ed. G.F.G. Stanley (Toronto and New York, 1955).

Northcott, Ruth, ed. *Astronomy in Canada, Yesterday, Today, and Tomorrow* (Toronto, 1967).

O'Brien, Charles F. 'Eozoön Canadense "The Dawn Animal of Canada"' *Isis 61* (1970), 206-23.

O'Brien, Charles F. *Sir William Dawson: A Life in Science and Religion* (Memoirs of the American Philosophical Society, 84). Philadelphia; American Philosophical Society, 1971.

Thomson, Don W. *Men and Meridians*, 3 vols. (Ottawa, 1966-1969).

D. SCIENCE AND THE GENERAL PUBLIC (Section 3)

Appleman, Philip, ed. *Darwin. A Norton Critical Edition* (New York, 1970).

Canada, Dept. of Agriculture. *Fifty Years of Progress on Dominion Experimental Farms, 1886-1936* (Ottawa, 1939).

Clow, A., and Clow, N.L. *The Chemical Revolution* (London, 1952).

Ellegård, Alvar. *Darwin and the General Reader* (Göteborg, 1958).

Fay, C.R. *Palace of Industry 1851: A Study of the Great Exhibition and Its Truth* (Cambridge, 1951).

Gillispie, Charles C. *Genesis and Geology. The impact of scientific discoveries upon religious beliefs in the decades before Darwin* (New York, 1959).

Irvine, William. *Apes, Angels and Victorians. Darwin, Huxley, and Evolution* (Cleveland and New York, 1959).

James, C.C. 'The First Agricultural Societies', *Queen's Quarterly*, 10 (1902), 218-23.

Literary and Historical Society of Quebec. *The Centenary Volume of the Literary and Historical Society of Quebec* (Quebec, 1924).

Lortie, Léon. 'Les Sciences à Montréal et à Québec au xixᵉ siècle', *L'Action Universitaire*, 1, n° 3 (1936).

Royal Society of Canada. *The Royal Society of Canada, 1882-1957* (Ottawa, 1958).

Wallace, W.S., ed. *The Royal Canadian Institute Centenary Volume* (Toronto, 1949).

White, A.D. *A History of the Warfare of Science with Theology in Christendom*, 2 vols (New York, 1960; reprint of first edition, 1896).

E. SCIENTIFIC EDUCATION AND RESEARCH (Section 4)

Bailey, L.W. 'Dr James Robb', *Bulletin of the Natural History Society of New Brunswick*, 16 (1898), 1-15.

Clark, A.L. *The First Fifty Years: A History of the Science Faculty at Queen's, 1893-1943* (Kingston, 1944).

Doern, G. Bruce. *Science and Politics in Canada* (Montreal, 1972).

Haber, L.F. *The Chemical Industry during the Nineteenth Century* (Oxford, 1958).

Jarrell, Richard A. 'Science Education at the University of New Brunswick in the Nineteenth Century', *Acadiensis* (1973), 55-79.

Senate Special Committee on Science Policy. *A Science Policy for Canada*, 3 vols (Ottawa, 1972-3).

Warrington, C.J.S., and Nicholls, R.V.V. *A History of Chemistry in Canada* (Toronto, 1949).

Winslow-Spragge, Lois. *Life and Letters of George Mercer Dawson, 1849-1901* (Privately printed Montreal, 1962).